# LOGIC SYNTHESIS USING SYNOPSYS®
# SECOND EDITION

**Trademark Information**

UNIX is a registered trademark of UNIX System Laboratories, Inc.
Cadence is a registered trademark of Cadence Design Systems, Inc.
Composer and Verilog are registered trademarks of Cadence Design Systems, Inc.
Xilinx is registered trademark of Xilinx, Inc.
PPR, XNF, Makebits, LCA, lca2xnf are registered trademarks of Xilinx, Inc.
Synopsys is a registered trademark of Synopsys, Inc.
Design Analyzer, Design Compiler, Test Compiler, VHDL Compiler, HDL Compiler, Library Compiler, Simulation Graphical Environment (SGE), VHDL System Simulator, charecterize, dont_touch, dont_use and dont_touch_network, Synthetic libraries, DC Expert, DC Professional, DesignTime, DesignWare, Designware Developer, Floorplan Manager, are registered trademarks of Synopsys, Inc.
lsi_10k is a registered trademark of LSI Logic Corporation.
Mentor and Falcon Framework are registered trademarks of Mentor Graphics Corporation.
ViewLogic is a registered trademark of ViewLogic Corporation.
Viewgen, Viewdraw, vl2edif2 are registered trademarks of ViewLogic Systems, Inc.
Sun Workstation and SPARC are registered trademarks of Sun Microsystems.
XACT, IOB, XNF and CLB are trademarks of Xilinx, Inc.

# LOGIC SYNTHESIS USING SYNOPSYS®
# SECOND EDITION

by

**Pran Kurup**
Cirrus Logic, Inc.

and

**Taher Abbasi**
Synopsys, Inc.

**KLUWER ACADEMIC PUBLISHERS**
**Boston / Dordrecht / London**

**Distributors for North America:**
Kluwer Academic Publishers
101 Philip Drive
Assinippi Park
Norwell, Massachusetts 02061 USA

**Distributors for all other countries:**
Kluwer Academic Publishers Group
Distribution Centre
Post Office Box 322
3300 AH Dordrecht, THE NETHERLANDS

*Consulting Editor: Jonathan Allen, Massachusetts Institute of Technology*

**Library of Congress Cataloging-in-Publication Data**

A C.I.P. Catalogue record for this book is available
from the Library of Congress.



*Printed on acid-free paper.*

Printed in the United States of America

*To Our Parents*

# *Table of Contents*

# *Foreword*

Advances in integrated circuit technology have given us the ability to design and manufacture large integrated circuits. However, these advances have increased the number of parameters that a designer must deal with in order to realize a high-quality, commercially viable design. Smaller geometries of semiconductor technologies that have enabled larger and faster designs have also significantly impacted the variables that affect timing in designs and the increased number of design alternatives that need to be considered. The limitations of schematic capture based design have become clearly evident with the increase in complexity of integrated circuits. In the fast changing world of integrated circuits, high-level design and logic synthesis have provided an answer to these limitations. High-level design methodology using hardware description languages (HDLs) such as VHDL and Verilog, has emerged as the primary means to capture functionality and deal with a range of design issues. It must be recognized that there is more to design capture using VHDL or Verilog than creating a textual equivalent of a schematic. Further, merely capturing functionality in a high-level language is hardly sufficient. The high-level description must match the required design specifications and must be synthesizable. Moreover, the HDL code must be well-structured, readable and modular to enable re-use with minimal effort.

The use of logic synthesis has made it possible to effectively translate designs captured in these high-level languages to designs optimized for area and speed. Besides enabling control over design parameters such as silicon real-estate and timing, logic synthesis tools facilitate the capture of designs in a parametrizable and re-usable form. Moreover, logic synthesis makes it possible to re-target a given design to new and emerging semiconductor technologies.

Present-day IC-design methodology based on high level design capture and the use of logic synthesis is far from a "push-button" solution. To realize a high quality design, the designer must simultaneously consider both the coding of the design and the requirements for logic synthesis. Some HDL constructs synthesize more effectively than others. HDL code which might simulate correctly need not necessarily be synthesizable. In short, there is a need to design for synthesis. Synthesis is an iterative process aimed at achieving design goals such as an optimal number of gates and a desired speed of operation. There are several "do's and don'ts" that a designer must be aware of when coding in HDL for synthesis.

The use of high-level design tools have facilitated effective sharing of re-usable designs and methodologies across several designs. However, an effective design methodology utilizing an optimal combination of design tools is a continuing challenge. While the design community constantly aspires to innovate and realize the best out of these tools, they simultaneously invest substantial efforts in developing the required expertise in their usage.

Realizing the full potential of synthesis requires an in-depth understanding of the synthesis process and the working of synthesis tools. This book by Pran Kurup and Taher Abbasi, helps the reader gain an in-depth understanding of high-level design capture and the logic synthesis process. The authors follow a practical approach to addressing the relevant issues in both VHDL and Verilog in conjunction with the Synopsys Design Compiler. The book presents several examples derived from real-world situations and should be useful to both aspiring and practicing IC design engineers.

*Dr. Suhas Patil*

*Chairman of the Board, Cirrus Logic, Inc.*

*Fremont, CA.*

# *Preface*

Schematic capture based approach to Integrated Circuit (IC) design was pioneered in the early eighties and was soon widely adopted. However, with major advances in Electronic Design Automation (EDA) has emerged a relatively automated means of IC design, based on hardware description languages (HDL). The transition to HDL-based designs has enabled a substantial increase in productivity with regard to "gates per engineer per day" when compared to schematic capture based designs. Further, constant improvements in fabrication technology have made possible ICs with over a million gates. At the center stage of this transition to HDL-based design lies "Logic Synthesis". Synthesis is essentially the process of transforming functionality which is initially described in HDL, to an optimized technology-specific netlist. The ever increasing demand to achieve highly complex, large gate count chips with a fast turnaround has propelled synthesis to the forefront of the HDL-based ASIC design process. Logic synthesis has since become a fundamental component of the ASIC design flow.

EDA tool vendors have been constantly upgrading their tool capabilities to realize the full potential of HDL-based design. As EDA vendors strive to improve their synthesis offerings, the whole process of synthesis has become increasingly complex. What was once considered as a mere means to "crunch out a netlist" from HDL source code, has become far more involved. The nuances of the synthesis process have begun to significantly impact not only the final netlist generated, but also a broad spectrum of issues ranging from HDL coding styles and design partitioning, to netlist integration and CAD methodology. Several design houses modify existing designs to develop improved higher performance designs. These designs are essentially derivatives of existing design databases. Synthesis facilitates easy re-use of existing technology-independent designs described in HDL.

The book is for anyone who hates reading manuals but would still like to learn logic synthesis as practiced in the real world. Synopsys *Design Compiler* the leading synthesis tool in the EDA marketplace, is the primary focus of this book. The contents of this book are specially organized to assist designers accustomed to schematic capture based design, develop the required expertise to effectively use the Synopsys

Design Compiler. Over 100 *"Classic Scenarios"* faced by designers when using the Design Compiler have been captured, discussed and solutions provided. These scenarios are based on both personal experiences and actual user queries. A general understanding of the problem solving techniques provided should help the reader debug similar and more complicated problems. In addition, several examples and dc_shell scripts (Design Compiler scripts) have also been provided.

The leading HDLs are clearly, Verilog and VHDL. Several books have been written to facilitate the speedy learning of VHDL and Verilog. *This book is not a VHDL or Verilog primer.* Basic knowledge of VHDL or Verilog is essential in understanding logic synthesis. A preliminary exposure to any commercially available synthesis tool, though not a pre-requisite, should help speed up the learning process. This book should specifically help the reader develop a better understanding of the synthesis design flow, optimization strategies using the Design Compiler, test insertion using the Test Compiler, commonly used interface formats such as EDIF and SDF, Links from the Design Compiler to Layout Tools, the FPGA synthesis process, design re-use in a synthesis-based design methodology and a conceptual introduction to behavioral synthesis. Examples in both VHDL and Verilog have been provided.

One of the most crucial issues in the ASIC design flow is interfacing between EDA tools. ASIC design houses which have migrated to a HDL-based design flow have adopted different strategies to incorporate synthesis into their existing design methodology. An effective ASIC design flow is largely dependent on seamlessly integrating a host of commercial and proprietary Electronic Design Automation (EDA) tools. Most design houses use three or more commercially available EDA tools. This is usually in addition to in-house proprietary CAD tools, not to mention the ubiquitous emacs/vi editors for minor hacks. From our experience of working with ASIC designers we believe that there is a greater need for a book exploring the basics: one which would facilitate a smooth transition to a synthesis-based design methodology. We have therefore, focused primarily on logic synthesis with the ASIC design flow in perspective.

The second edition covers several new and emerging areas in addition to improvements in the presentation and contents in all chapters from the first edition. With the rapid shrinking of process geometries it is becoming increasingly important that "phyical" phenomenon like clusters and wire loads be considered during the synthesis phase. The increasing demand for FPGAs has warranted a greater focus on FPGA synthesis tools and methodology. Finally, Behavioral synthesis, the move to designing at a higher level of abstraction than RTL, is fast becoming a reality. All these factors have prompted us to include separate chapters in the second edition to cover Links to Layout, FPGA Synthesis and Behavioral synthesis, respectively.

The book has also been written with the CAD engineer in mind. A clear understanding of the synthesis tool concepts, its capabilities and the related CAD issues should help the CAD engineer formulate an effective synthesis-based ASIC design methodology. The intent is also to assist design teams to better incorporate and effectively integrate synthesis with their existing in-house design methodology and CAD tools. We anticipate that this book will provide answers to several issues that confront both ASIC designers and CAD engineers. Feedback and comments on this book are most welcome, and can be sent to taher@bytekinc.com.

## *Overview of the Chapters*

Chapter 1 provides an overview of the synthesis based ASIC design flow. In subsequent chapters, each of the steps in the flow have been dealt with in greater detail. On reading chapter 1, the reader should be able to directly proceed to specific issues discussed in subsequent chapters of the book, if desired.

Chapter 2 deals with the very first stage in the synthesis process - HDL coding. Several recommendations for writing synthesizable HDL code have been discussed in this chapter. Coding for finite state machines is discussed in great detail with examples in both VHDL and Verilog.

Chapter 3 covers the simulation steps, i.e., both functional or behavioral simulation of the HDL and the post-synthesis or gate-level simulation. This chapter has been included, primarily, to help the reader understand the design flow better. An example VHDL code of TAP controller is provided along with its testbench. The steps to perform behavioral simulation, and gate level simulation after synthesis have been discussed. The simulator used here is the Synopsys *VHDL System Simulator (VSS)*. However, discussion on VSS is beyond the scope of this book.

Chapter 4 and 5 are all about constraining designs to achieve the optimal design. How to get the best out of the Design Compiler? Chapter 4 begins with an introduction to synthesis covering optimization constraints, design rule constraints and Design Compiler timing reports. This is followed by a description of commonly used Design Compiler terminology. The rest of the chapter covers, steps in optimizing designs, and general guidelines for synthesis. Discussion of over 25 classic scenarios is also provided. Chapter 5 describes FSM synthesis and other useful synthesis capabilities such as fixing timing violations and technology translation. Several examples with scripts have been provided.

Chapter 6 is devoted to Links to Layout. What are the mechanisms available for links from front-end tools like Design Compiler to backend tools such as floor-planning and place and route tools? How does one perform in-place optimization? This chapter addresses these and similar issues relating to links between Synopsys logic synthesis tools and back-end tools using the Synopsys *Floorplan Manager*.

Chapter 7 discusses FPGA synthesis using the Synopsys *FPGA Compiler*. The target FPGA library described in this chapter is the Xilinx XC4000 family of FPGAs. FPGA synthesis has gained in significance thanks to the increasing market for FPGAs driven primarily by their rapidly increasing speed and density. This chapter discusses the synthesis flow for targetting FPGAs.

Chapter 8 covers the concepts of testability and how they can be incorporated into the synthesis design flow. This chapter focuses on testability using the Synopsys *Test Compiler (TC)*. Guidelines to be followed when using the TC are outlined. Again, several classic scenarios have also been discussed.

Chapter 9 is devoted to interfaces. This chapter provides a basic discussion on EDIF generated by the Design Compiler. EDIF constructs are explained via an example EDIF schematic generated by the Design Compiler. This chapter also includes discussion on other commonly used interface fomats such as SDF and PDEF.

Chapter 10 discusses the design re-use approach using synthesis which facilitates the design of high-performance, high-integration chips with a fast turnaround. The DesignWare concept is presented and the methodology for building DesignWare libraries using the *DesignWare Developer* described. Techniques to map HDL to complex technology library cells or pre-defined structures using the Design Compiler are discussed in detail with examples.

Chapter 11 introduces the reader to Behavioral Synthesis. The EDA tool used is the Synopsys *Behavioral Compiler*. What is Behavioral synthesis? Is behavioral synthesis right for you? What does a transition to Behavioral synthesis entail? These issues are addressed in this introductory chapter on behavioral synthesis using a simple example.

Work on this book has spanned several software releases of Synopsys tools. Small changes in commands in different versions are likely, but forward compatability can be assured in most cases. At the end of each chapter a "Recommended Reading" list has been provided to help the reader access the appropriate Synopsys documentation, if required. Finally, the Appendix provides sample dc_shell scripts and a brief introduction on Synopsys On-line Documentation (iview). These scripts have been made as general purpose as possible. If you have mastered the material in the above chapters, writing scripts from scratch should be as simple as modifying these scripts to suit your needs!

# *Acknowledgments*

We wish to thank *Synopsys, Inc.* for providing us the opportunity to learn logic synthesis. The inspiration for this book came from our extensive experience of working with customers as Applications Engineers for Synopsys, Inc. Our thanks to all Synopsys users whom we have had an opportunity to interact with at different times. We wish to express our gratitude to Robert Dahlberg of Synopsys, Inc. Though this book was an independent initiative on our part, it was his "Help the customers help themselves" concept, that first encouraged us to consider this project. We are deeply indebted to Dr. Suhas Patil, Chairman, Cirrus Logic Inc., for reviewing this book and contributing the foreword.

We also extend our appreciation to Michael Albrecht, Amir Mottaez, Pradeep Fernandes and Rob Maffit of Synopsys, Inc., Shridhar Mukund, Chetan Patil and Shekhar Patkar of Cirrus Logic, Inc., Dr. J. Anandkumar of Hughes Network Systems, Yatin Trivedi of Seva Technologies, Prasad Saggurti of Sun Microsystems, Prof. Sudhakar Yellamanchalli of Georgia Institute of Technology and Prof. Jon Allen of Massachusetts Institute of Technology (MIT), for their comments and feedback. We wish to thank H.Ravindra, John Geldman of Cirrus Logic, Inc., Alain Labat, Jack Warecki and Vito Mazzarino of Synopsys, Inc., for their support and encouragement.

Carl Harris at Kluwer Academic Publishers deserves special mention for his unstinted patience, co-operation and guidance. Special thanks to Lachmi Khemlani of University of California at Berkeley for the design of the cover. We also wish to thank Tina Jalalian for her utmost co-operation and the hard work in formatting, editing and preparation of the final manuscript.

*Pran Kurup & Taher Abbasi*

# About The Authors

Pran Kurup is a Project Manager at Cirrus Logic, Inc. in the R&D department. Cirrus Logic, Inc. is a leading supplier of advanced integrated circuits for desktop and portable computing, telecommunications and consumer electronics markets. At Cirrus, Pran has been actively involved in ASIC design and development of methodologies using logic synthesis. He is also responsible for management of project accounts within the R&D department of Cirrus Logic and is extensively involved in the evaluation of CAD tools. Prior to this he was an Applications Engineer at Synopsys Inc. for a period of two years. He has worked with several Synopsys customers supporting Synopsys tools, particularly the Design Compiler and interfaces to other EDA vendor tools. Pran completed his undergraduate in Electrical Engineering from the Indian Institute of Technology, Kharagpur, in India and his Masters in Computer Engineering from the University of Miami, Coral Gables, in Florida. He has written articles for the ASIC & EDA magazine, EDN-Asia magazine, the Synopsys Methodology Notes and application notes at Cirrus Logic, Inc.

Taher Abbasi is co-founder and President of ByteK Designs Inc., a design solutions company based in Palo Alto, California specializing in ASIC design, methodology development, training and contracting services. Prior to this, he was a Senior Design Applications Engineer at Synopsys, Inc. in Mountain View, California. He has worked with several Synopsys customers supporting the entire range of Synopsys tools, with special emphasis on the Test Compiler, the Behavioral Compiler, the Design Compiler, VHDL Compiler and Designware. Taher has contributed articles to "Impact" the Synopsys quarterly newsletter and to EDN-Asia magazine. His current interests include the Synopsys Behavioral Compiler and the development of methodologies using high-level design tools. He has been on consulting assignments for Synopsys at several leading system and semiconductor companies. Taher also teaches Test Synthesis courses for Synopsys customers and along with Pran is a co-instructor of the "Advanced Logic Synthesis Using Synopsys" class offered at the University of California, Santa Cruz Extension and California State University, Northridge. He completed his Masters in Computer Engineering from California State University, Northridge, and his undergraduate in Electronics Engineering from Bombay University, India.

CHAPTER 1

# *High-Level Design Methodology Overview*

Major advances in fabrication technology have made possible high-integration, large gate count ASICs. Hardware description languages and logic synthesis have had a significant impact on the design process of these ASICs. With the adoption of HDL-based design, there has emerged a high-level design flow based on synthesis.

The most commonly used HDLs today are VHDL and Verilog. The desired functionality of a design is first captured in HDL code, usually Verilog or VHDL. This step is complex, particularly for IC designers who are accustomed to schematic capture tools. This is further compounded by the fact that this code then must be synthesized into an optimal design which meets the functional requirements of the initial specification.

In this chapter, the entire synthesis-based ASIC design flow and methodology are discussed. Each of the steps in this process is then described in greater detail in subsequent chapters. This chapter is intended primarily to emphasize the ASIC design flow involving synthesis.

## 1.1 ASIC Design Flow Using Synthesis

The synthesis based ASIC design flow includes the following steps:

1. Functional specification of the design
2. HDL coding in VHDL/Verilog RTL
3. RTL/behavioral or functional simulation of the HDL
4. Logic synthesis
5. Test insertion and ATPG
6. Post-synthesis or gate-level simulation
7. Floorplanning / place and route

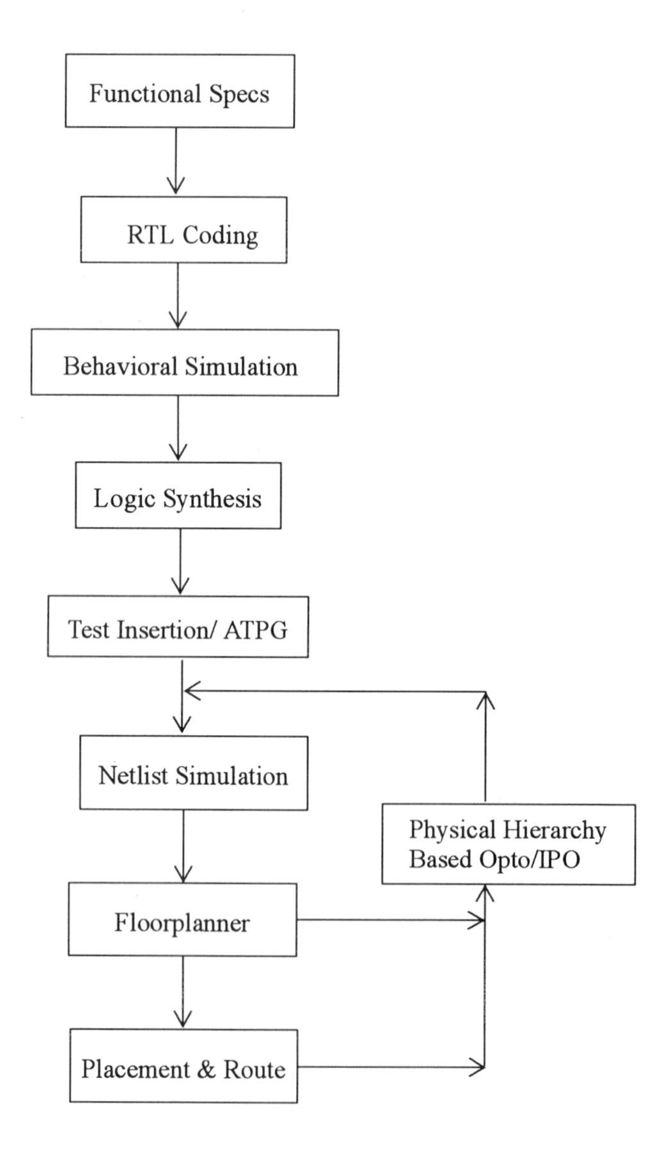


**Figure 1.1 Synthesis Based ASIC Design Flow**

The above seven steps are iterative as shown in Figure 1.1. For example, on performing a functional simulation of the HDL source code, one might find that the code does not exactly match the desired functional behavior. In such cases, one must return to modify the source code. Also, after synthesis, it is possible that the netlist does not meet the timing requirements of the clock. This implies that one must either modify the source HDL, or attempt alternate synthesis strategies, or in more unfortunate, time-crunch situations, "hack the netlist". Similarly, after performing place and route, it is required to back annotate delay values to incorporate real-world delays. This is followed by *in place optimization (IPO)* of the netlist to meet routing delays. Clearly, the steps involved in synthesis-based ASIC design are iterative.

The functional specification is always the first step in an ASIC design process. Designers in particular, are extremely familiar with formulating the specifications of a design. Most often the design specification (on paper) is followed by a block level diagram of the entire ASIC. The block level diagram of the ASIC is usually done using graphical design entry tools such as the *Simulation Graphical Environment (SGE) (Synopsys)* or *Composer (Cadence)*.

After a block level schematic capture of the design, the next step involves HDL coding. The style of HDL coding often has a direct impact on the results the synthesis tool delivers. A sound knowledge of the working of the synthesis tool will help the designer write synthesizable code better. For example, one common problem arises due to partitioning of designs. Typically, designers partition designs based on functionality. During the integration of different modules in synthesis, one might find a large amount of logic in the critical path. This critical path most often traverses several hierarchical boundaries. In a typical design team scenario, these blocks are usually designed by different engineers, thereby compounding the problem.

*Logic synthesis provides the best results when the critical path lies in one hierarchical block as opposed to traversing multiple hierarchical blocks.* In such situations, it is often required to modify the hierarchy in the source HDL code and re-optimize the design or modify the design hierarchy through Synopsys tool specific scripts. This is just one classic example of how one has to adapt the design process to get the best out of synthesis. Another coding tip is to *ensure that all outputs of sub-blocks/modules are registered outputs*. In synthesis, this helps to estimate the input delays and helps avoid intricate time budgeting. Also, it is recommended that logic blocks such as ROMs and random logic each be grouped into separate hierarchical blocks. We defer the details for coding tips to Chapter 2.

### 1.1.1 HDL Coding

HDL code can be behavioral or RTL. In the synthesis domain, the latter is usually considered to be the synthesizable form of HDL code. Since the focus of this book is on logic synthesis, all examples discussed are in synthesizable RTL code. As a preliminary introduction to HDL coding and synthesis let us consider the following simple examples. Four examples of HDL code which infer on synthesis a D-flip flop, a latch, an AND gate, and a multiplexer (referred to as *mux* throughout this book) respectively are discussed.

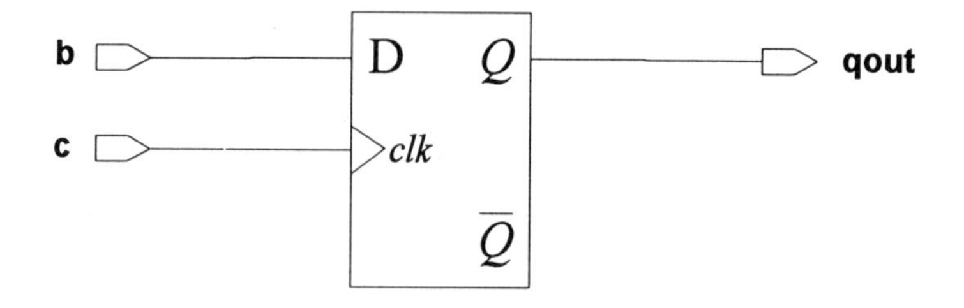

**Figure 1.2 D-Flip Flop**

## Example 1.1 Code for D Flip Flop

### VHDL Code

```
entity flipflop is
port (b,c : in bit;
      qout : out bit);
end flipflop;
architecture test of flipflop is
begin
process
begin
       wait until c'event and c = '1' ;
       qout <= b ;
end process;
end test;
```

### Verilog Code

```
module flipflop(b,c,qout);
input b,c;
output qout;
reg qout;
always@ (posedge c)
      qout <= b;
endmodule
```

Example 1.1 shows VHDL and Verilog code which when synthesized infers a positive edge triggered flip flop. If one desires a negative edge triggered flip flop, the obvious change to make would be to replace the posedge declaration in the code with negedge in the Verilog code (or clock='1' by clock='0' in VHDL). While this might appear to be an obvious solution, inferring a negative edge triggered flip flop is largely dependent on an appropriate negative edge triggered flip-flop being available in the technology library. Alternately, one can always instantiate a negative edge triggered flip flop (using structural HDL code) from the ASIC vendor library to circumvent the problem. However, this would mean that the HDL is bound to a certain technology library.

In a larger design, which has both positive and negative edge triggered flip flops, it is recommended that *all the positive edge triggered flip flops be grouped into a separate level of hierarchy and all the negative edge triggered flip flops be grouped into another level of hierarchy*. This makes the debug process and timing analysis during synthesis simpler.

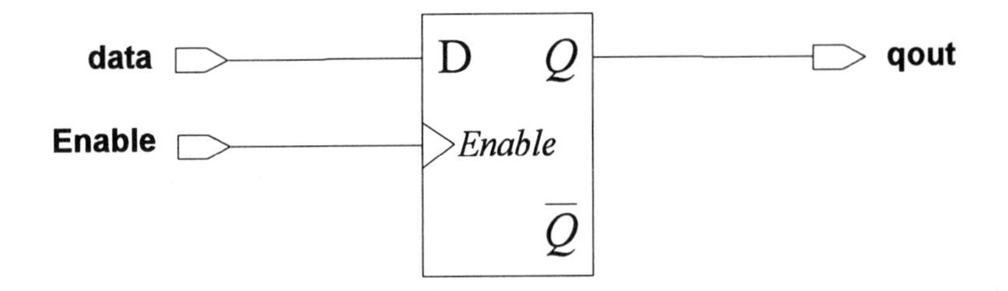


**Figure 1.3 D-latch**

## Example 1.2 Code for Latch

### VHDL Code

```
entity latch is
port (data,enable : in bit;
        qout : out bit);
end latch;
architecture test of latch is
begin
process (data, enable)
```

```
begin
   if ( enable = '1') then
        qout <= data ;
   end if;
end process;
end test;
```

**Verilog Code**

```
module latch(enable,data,qout);
input enable,data;
output qout;
reg qout;
always @ (enable or data)
      if (enable == 1)
        qout = data;
endmodule
```

Example 1.2 shows the Verilog and VHDL code which when synthesized infers a latch. Notice that the if statement does not have an else clause specified. The synthesis tool interprets the absence of an else clause to mean, "when if condition not satisfied, maintain previous value". Hence a latch is inferred in this example.

## Example 1.3 Initialized Signal to Prevent Latch Inference

**VHDL Code**

```
entity sample is
port (b,enable : in bit;
      d : out bit);
end sample;
architecture test of sample is
begin
process (enable)
begin
    d<= '0' ; -- Initialized signal d to 0
    if ( enable = '1') then
        d <= b ;
    end if;
end process;
```

```
end test;
```

**Verilog Code**

```
module sample(b, enable, d);
input b, enable;
output d;
reg d;
always @(enable or b) begin
  d = 1'b0;  // initialized signal to prevent latch inference
  if (enable == 1'b1)
      d = b;
end
endmodule
```

Example 1.3 is similar to Example 1.2, except for the initialized signal d. Notice that signal d is initialized to a value 0. This should prevent a latch from being inferred. Most often in larger designs, unwanted latches are inferred when signals are not initialized. Hence, it is always advisable to *initialize signals* and then assign them values depending on certain conditions being satisfied, as in Example 1.3. This should prevent unwanted latches from being inferred.

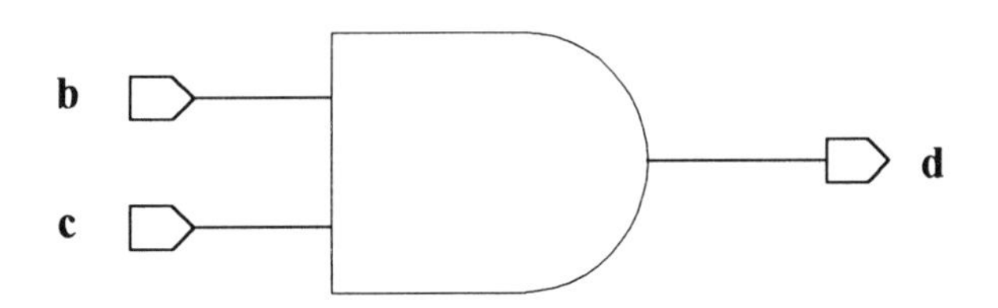


**Figure 1.4 AND Gate**

## Example 1.4 Code for AND Gate

**VHDL Code**

```
entity AND2 is
port (b,c : in bit;
      d : out bit);
end AND2;
```

```
architecture test of AND2 is
begin
process
begin
    if ( c = '1') then
        d <= b ;
    else
        d <= '0' ;
    end if;
end process;
end test;
```

**Verilog Code**

```
module and2(c,b,d);
input c,b;
output d;
reg out;
always@ (c or b)
  if (c == 1)
       d = b;
  else
       d = 0;
endmodule
```

Example 1.2 and Example 1.4 are similar but result in different logic from being inferred. Example 1.2 differs from Example 1.4 in that it does not have an else clause. Example 1.4 infers an AND gate when synthesized. In this case, the HDL code exactly matches the logic inferred. The above example has been shown to describe how similar code can result in different logic from being inferred. In general, to infer an AND gate, the recommended coding style for synthesis is using (out = a&b;) instead of the if statement. In this example, the coding style with the if statement has been used, to show how a very small difference in HDL code can drastically impact the logic inferred.

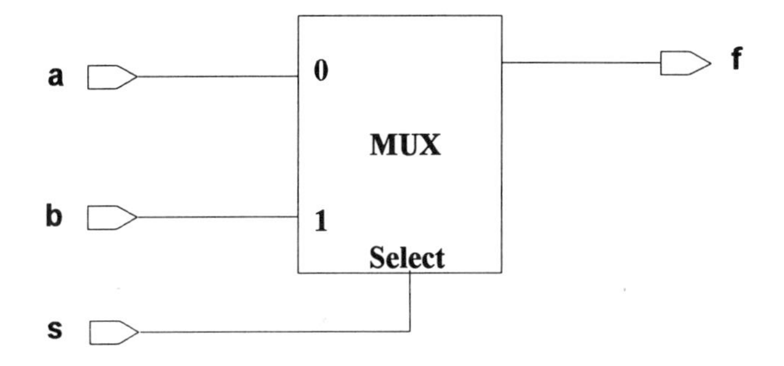


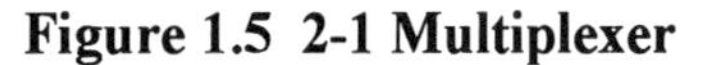

**Figure 1.5 2-1 Multiplexer**

### Example 1.5 Code for MUX

**VHDL Code**

```
entity mux is
port (a, b, s : in bit;
        f : out bit);
end mux;
architecture test of mux is
begin
process (a, b , s)
begin
  case s is
    when '0' => f<= a;
    when '1' => f <= b;
  end case;
end process;
end test;
```

**Verilog Code**

```
module mux(a,b,s,f);
input a,b, s;
output f ;
reg f;
always@ (a or b or s)
```

```
begin
    case (s)
        1'b0 : f = b;
        1'b1 : f = a;
    endcase
end
endmodule
```

Example 1.5 shows code which on synthesis infers a mux. However, it is not uncommon to find that the tool infers gates which have the same functionality as a mux. In the case of larger muxes like 8:1 or 4:1 muxes, it is almost always better to instantiate them. This is because the synthesis tool most often does not map to larger muxes. However, it is possible to force DC to map to muxes in the technology library using Synopsys tool specific directives (such as map_to_entity pragma) or through function calls. Another alternative to force the synthesis tool to map to muxes is by instantiating muxes from the Synopsys generic library (gtech.db) and assigning the map_only attribute (a Synopsys tool specific attribute) to it. Both these techniques are discussed in greater detail in Chapter 10.

Chapter 2 discusses HDL coding in greater detail with several examples. The issues discussed above have all been covered in greater depth.

### 1.1.2 RTL/Behavioral and Gate-Level Simulation

After the design has been captured in HDL, it is essential to verify that the code matches the required functionality, prior to synthesis. We call this step the *pre-synthesis behavioral simulation* of the HDL. This can be performed by simply assigning specific values to input signals, performing simulation runs and viewing the waveforms in a graphical simulation tool. An alternative is to write a testbench The testbench can be considered as an HDL block whose outputs provide the stimuli for the design to be simulated. In general, it is recommended that one write a testbench for simulation to simplify the post-synthesis simulation step. The same testbench can then be used for post synthesis simulation and the results of the two compared.

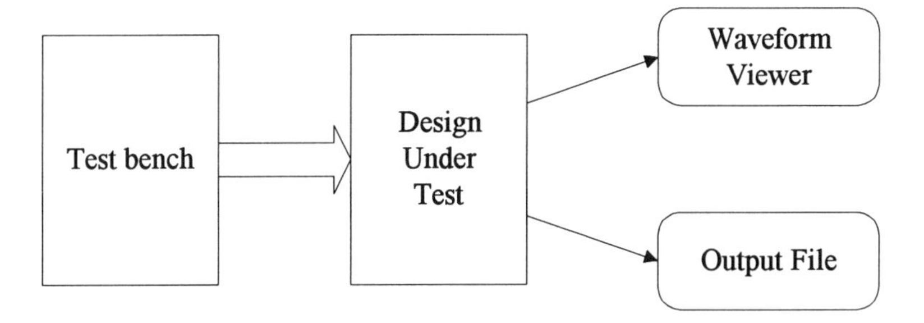


**Figure 1.6 Simulation Flow using a Testbench**

There are two possible ways of simulating a design using a testbench as shown in Figure 1.6. One can write a testbench where all the stimuli for the different signals are provided in the HDL code. Then one would use a graphical front end of a simulation tool to view the waveforms. It is also possible to provide the stimuli for the different signals and pipe the outputs to a file. If one has another file of expected results, one can quite easily compare the two files to ensure that the two match. Both these techniques have been discussed in chapter 3 with examples.

To simulate a synthesized gate-level netlist, VHDL simulation models of the technology library cells are required. These can be of four kinds - unit delay structural model (UDSM), full-timing structural model (FTSM), full-timing behavioral model (FTBM), or full-timing optimized gate-level simulation (FTGS). While UDSM and FTSM are used for functional verification, the FTBM is used for accurate, detailed timing verification and FTGS library for fast, sign-off-quality timing verification.

- Unit Delay Structural Model (UDSM)

  In this model of a technology library, all combinational cells have a rise/fall delay of 1ns, while all sequential cells have a rise/fall delay of 2 ns.

- Full-Timing Structural Model (FTSM)

  This model includes transport wire delays and pin-to-pin delays on a zero delay functional network. Timing constraint violations are reported as warning messages.

- Full-Timing Behavioral Model (FTBM)

  This delay model is used for detailed timing verification. Transport wire delays and pin-to-pin delays are included in this delay model.

- Full-Timing optimized Gate-level Simulation (FTGS)

  These models include transport wire delays and pin-to-pin delays in the delay model. In addition to warning messages, the Simulator can schedule X output values for timing constraint violations and circuit hazards. One can use the FTGS library for fast, sign-off-quality timing verification.

If one has the ASIC vendor library (the Synopsys .db file) for synthesis and the Synopsys Library Compiler the above VHDL simulation models can be automatically created using the liban utility. The Library Compiler does not, however, write out Verilog simulation models for technology library cells. The liban utility creates two files from the ASIC vendor library in Synopsys db (database) format. For example, given the technology library mylib.db, the liban utility generates an encrypted VHDL (mylib.vhd.E) containing simulation models with timing delays, and a VHDL package (mylib_components.vhd) containing the component declarations for all the cells in the ASIC vendor library.

*Note: The encrypted file works only with the Synopsys VHDL System Simulator(VSS).*

If one has the source library file (.lib file), one can write out VHDL models using the following dc_shell command. One can control the type of VHDL model written out (that is, UDSM, FTBM, FTSM or FTGS) by setting the dc_shell variable vhdllib_architecture.

```
write_lib -f vhdl
```

Behavioral simulation and examples of VHDL simulation models are provided in Chapter 3.

### 1.1.3 Logic Synthesis

For a designer familiar with ASIC design using schematic capture, the task of understanding synthesis is further compounded by synthesis world nomenclature. Synthesis, as referred to in present-day IC design, can be broadly divided into *logic synthesis* and *high-level synthesis*. High-level synthesis is closer to what some refer to as *behavioral synthesis*. High-level synthesis involves synthesis of logic from behavioral descriptions. Synopsys *Behavioral Compiler* is specially targeted towards behavioral synthesis. Logic synthesis on the other hand synthesizes logic from register transfer level (RTL) descriptions. In the Synopsys domain, DC capabilities such as arithmetic optimization, implementation selection, resource sharing, in place optimization and critical path re-synthesis are referred to as *high level optimization*. High level optimization must not be confused with behavioral synthesis. In this book, we deal with logic synthesis in greater detail.

The logic synthesis process consists of two steps - translation and optimization. Translation involves transforming a HDL (RTL) description to gates, while optimization involves selecting the optimal combination of ASIC technology library cells to achieve the required functionality.

### 1.1.4 Design For Testability

With the increasing complexity of ASIC designs, the cost of testing designs has become a fairly substantial component of the overall manufacturing and maintenance cost. The cost of testing is an outcome of several cost components such as test generation cost, testing time, automatic test equipment cost and so on. *Design for testability (DFT)* techniques have been used in recent years to reduce the cost of testing by defining testability criteria early in the design cycle.

The most popular DFT technique in ASIC design is the *Scan Design Technique*. Scan techniques involve replacing sequential elements in the design with equivalent scan cells. There exist different styles of scan cells. Based on the *scan style* selected the design is required to meet certain design rules. The most commonly used scan style is the multiplexed flip flop.

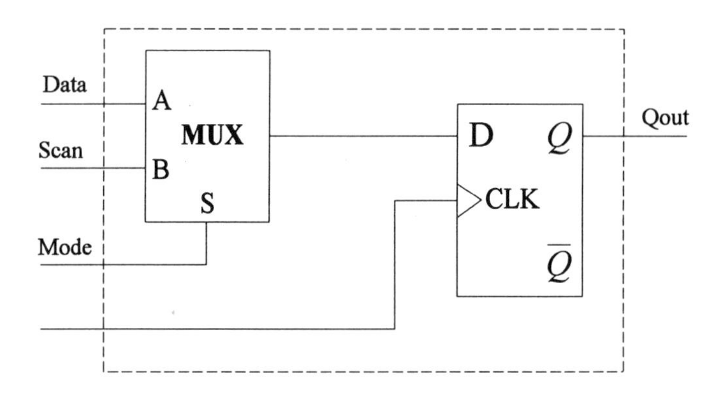


**Figure 1.7 Muxed Scan Flip Flop**

A muxed scan flip-flop, as the name indicates, consists of a mux and a flip-flop. The output of the mux drives the data input of the flip-flop and the select input is controlled by the test mode pin; the inputs to the mux are the data input and the test input as shown in the Figure 1.7. Sequential cells are replaced by scan equivalents to achieve the primary requirement of testability, namely, observability and controllability. A scan cell has two modes of operation, the *normal mode* and the *test mode*. During the normal mode it behaves like the sequential cell it replaces, but in the test mode the scan input is loaded into the scan cell on the active edge of the clock transition.

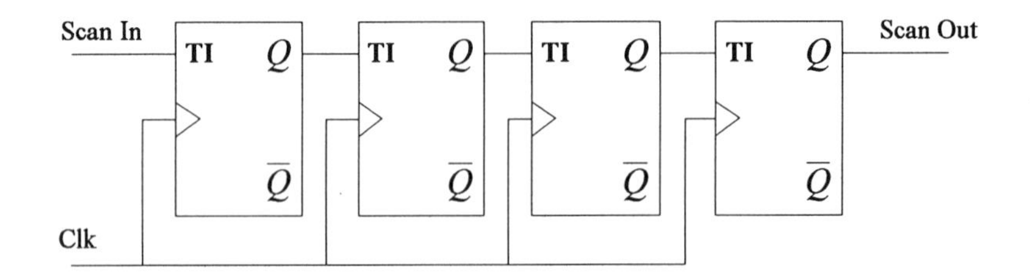


**Figure 1.8 Scan Chain Connected to Form a Shift Register**

Controllability and observability can be achieved when all the scan cells are connected to form a shift register in the test mode of operation as shown in Figure 1.8. While on-chip testability is being increasingly adopted in ASIC design, the extent to which it is implemented is often dependent on the target application, the area overhead, and the required fault coverage. Hence, within the scope of scan design, there exist two possible test methodologies:

1. Full scan
2. Partial scan

If all the sequential cells are replaced by scan cells, then it is called a *Full Scan* test methodology. In this case, the *Automatic Test Pattern Generation (ATPG)* algorithm is combinational. In a *Partial Scan* test methodology only some of the sequential cells are replaced by scan cells. In this case, the ATPG algorithm is sequential. Also, for partial scan it is required that the decision be made regarding which sequential cells are to be replaced by scan cells. This critical decision is usually based on the desired fault coverage and the available silicon area.

Once the test methodology has been specified as either partial or full scan, and the scan style declared, the Synopsys Test Compiler (referred to as TC for short, throughout this book) automatically replaces the sequential cells by scan cells. This replacement of sequential cells with scan equivalents occurs if the target technology library has scan equivalent cells of the required scan style, and provided no test design rule violations exist. It is possible that some flip flops in the design cannot be replaced by a scan equivalent due to design rule violations or user specified controls. In such cases, if full scan is the adopted methodology, then all the non-scan cells are interpreted as *black boxes*. This implies that all the faults associated with the black-box are untestable and hence results in reduced overall fault coverage.

After scan insertion, TC generates the test patterns for the design and the fault coverage achieved is calculated. The fault coverage is calculated as a percentage of the testable faults upon the total number of possible stuck-at-0 and stuck-at-1 faults. In general, full scan designs tend to achieve a higher fault coverage than partial scan designs. The amount of fault coverage for a partial scan design is related to the number of scannable registers in the design. After ATPG the test vectors must be formatted in one of the formats supported by the simulator on which the test vectors are to be simulated. Methodology issues and tips when using TC are discussed in Chapter 8.

### 1.1.5 Design Re-Use

Several design houses rely on re-use of large blocks of designs when building newer revisions of existing chips. In some cases, it might involve re-targeting an existing design to a new technology library. For others, it might involve minor tweaks to existing designs. In general, the strategy used to realize these changes has a significant impact on the turnaround time. Design re-use is an effective means to achieve fast turnaround on complex designs. A clear advantage in time is gained by re-use of designs from a library of parts rather than designing from scratch. The upcoming generation of complex systems will require a widespread availability of re-usable parts. Synthesis provides a very efficient and flexible mechanism to build a library of re-usable components. This is essentially the Synopsys DesignWare (DW) approach.

#### DesignWare Component Libraries

Synopsys provides DesignWare component libraries with existing re-usable parts. Examples of DesignWare libraries provided by Synopsys include, Standard, ALU, Advanced-Math, Sequential, Data Integrity, Control Logic and DSP families. The number of these libraries is certain to increase with subsequent versions of the software. In addition, there exists the generic GTECH library. When a source HDL is read into DC, the design is converted to a netlist of GTECH components and inferred DW parts. The GTECH library, like the DW libraries, is a technology independent library that helps users develop technology independent parts. The GTECH library called gtech.db contains common logic elements such as basic logic gates and flip flops. In addition, the gtech.db also contains a half adder and a full adder.

The DW libraries contain relatively more complex cells such as adders, subtracters, shifters, FIFOs, counters, comparators and decoders. These parts are *parametrizable, synthesizable, testable* and *technology independent* making them easily usable in a HDL design flow. Moreover, these parts have simulation models provided with the libraries, thus substantially improving the time required for both HDL coding and functional simulation. These parts serve as off-the-shelf design modules which can be used as functional sub-blocks in larger designs. Moreover, the synthesis tool ensures that when DW parts are used, high-level optimization features such as implementation selection, arithmetic optimization and resource sharing are automatically turned on.

DW components are flexible parts but *not* customized for particular designs. For example, if one requires a special feature in a FIFO component that is not found in the FIFO available with the DesignWare library, one cannot use this part. But in general, these parts are created with features such that they meet the requirements of most designs. Synopsys provides VHDL simulation models for all DW components and Verilog models for most DW components. The DW databook provides the latest list of available models.

DW also provides the capability for users to create their own DW libraries. In addition to re-use, these user built libraries provide a very effective mechanism to infer structures that DC would not normally map to via compile. This is very useful, especially when dealing with complex datapath structures. The concept of design re-use and parts available in the DW library are discussed in greater detail in Chapter 10.

## Designing with DesignWare Components

Consider the case of a Universal Asynchronous Receiver Transmitter (UART). The UART is used in almost all designs which involve serial to parallel and parallel to serial data conversion, and transmitting and receiving of data. A UART can be used in a number of designs provided it is sufficiently flexible.

Figure 1.9 shows a block diagram of the hierarchical overview of the UART. The UART requires several registers capable of performing shift operations, several counters, decoders and FIFOs. The Synopsys DesignWare libraries have pre-existing components namely, DW03_UPDN_CTR (up down counter), DW03_FIFO_S_DF (FIFO), DW01_DECODE (a decoder), DW03_SHFT_REG (shift register) which can be used in designing a UART.

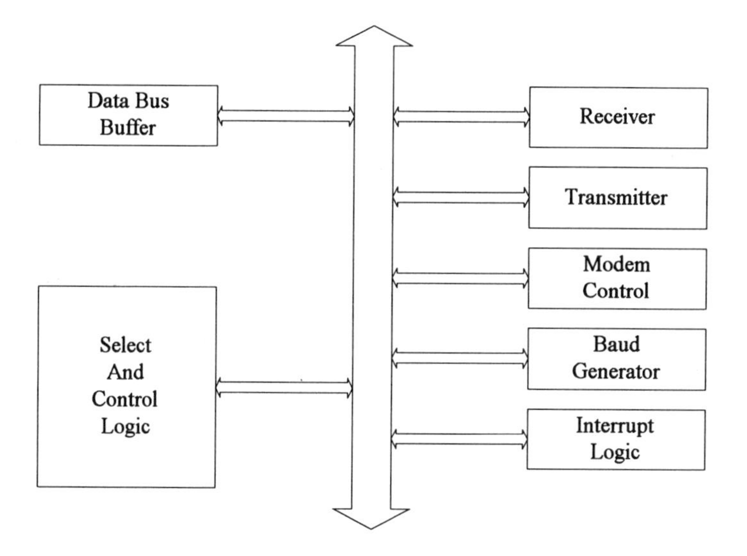


**Figure 1.9 Hierarchical Overview of UART**

Figure 1.10 shows the transmitter sub-block of the UART. The transmitter holding register can be built using the DW03_FIFO_S_DF. The DW03_DECODE can be used in the decode logic and the shift register (DW03_SHFT_REG) in the transmitter shift register block. The transmitter FSM is the only additional logic that is required to be designed.

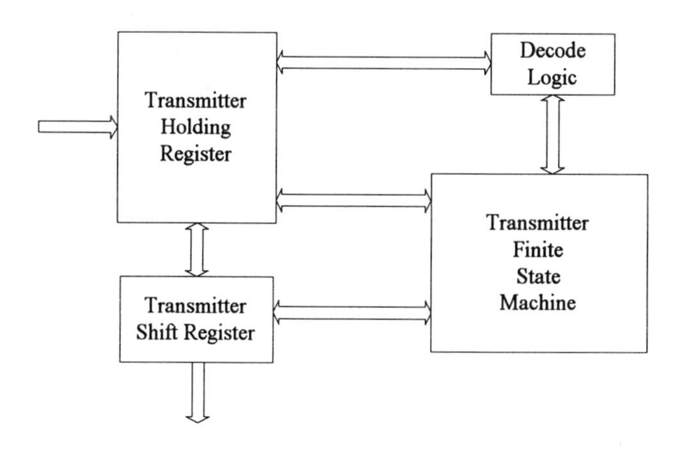


**Figure 1.10 Transmitter Block of UART**

### 1.1.6 FPGA Synthesis

Field Programmable Gate Arrays (FPGA) have emerged as an effective means to implement logic circuits at a relatively low cost with a fast turnaround. In other words, FPGAs have low prototype costs and shorter production times. An FPGA is a user-programmable integrated circuit, consisting of a set of logic blocks that can be interconnected using routing resources. The interconnect consists of segments of wires of various lengths and programmable switches that connect the logic blocks to the wire segments or wire segments to one another. Most importantly, the FPGA can be configured by the end-user, without the use of an expensive integrated circuit fabrication facility.

In recent times, FPGAs have rapidly gained in acceptance, driven by the increasing need for hardware *emulation*. Emulation is a technique of using programmable hardware as a physical prototype of custom and semi-custom ICs prior to silicon fabrication for chip and system verification. These developments along with rapid growth of FPGA technology (for example, density of FPGAs) has resulted in the need for synthesis tools with specific algorithms to exploit the architectural resources available on the FPGA, namely FPGA Synthesis tools. Examples of FPGA synthesis tools include FPGA Compiler for Synopsys and the Galileo tool suite from Exemplar. The key to FPGA synthesis lies in mapping the HDL description to the logic blocks available on an given FPGA. Chapter 7 provides a detail description of the FPGA synthesis design flow with specific reference to the Xilinx XC4000 family of FPGAs.

### 1.1.7 Links to Layout

With the advent of deep sub-micron technologies, interconnect or net delays have become a significant component of the overall delays. In other words, while gate delays decrease, the wire delays increase due to effects such as lateral capacitance, fringe capacitance and overlap capacitance. In the synthesis domain, it has become critical that "physical" information such as accurate wire loads and physical hierarchy information be taken into consideration upfront during the synthesis process. This requires effective links from the logic synthesis tool to backend tools such as floorplanning and place and route. The Synopsys Floorplan Manager(FPM) facilitates effective transfer of critical information to and from the backend tools to the Synopsys DC. The common formats for transfer of data to and from DC include, Standard Delay Format(SDF), Physical Data Exchange Format (PDEF) and Synopsys set_load script. Chapter 6 describes the links to layout mechanism and the design flow using the Synopsys Floorplan Manager in greater detail.

### 1.1.8 Behavioral Synthesis

Logic synthesis involves translation and optimization of an HDL description to a technology specific netlist. Behavioral synthesis on the other hand, involves synthesis at a higher level of abstraction - algorithms described in HDL as opposed to conventional RTL descriptions. In other words, functionality is described by which operations must occur as opposed to how they are implemented in hardware. Further, behavioral synthesis facilitates architectural exploration, in other words no pre-defined architecture as in logic synthesis. While logic synthesis optimizes for area and timing, behavioral synthesis adds another dimension to optimization namely, latency. In general, behavioral synthesis is suited to designs with complex data flow or I/O operations, and several memory accesses. It is meant for fully synchronous designs and implies faster simulation. A tight integration is required between Behavioral synthesis and Logic synthesis tools, since the output of behavioral synthesis is an RTL description which is then synthesized to gates using logic synthesis. Chapter 11 provides an introduction to behavioral synthesis using the *Synopsys Behavioral Compiler*.

## 1.2 Design Compiler Basics

This section is meant to familiarize the reader with frequently used synthesis terminology. The terms discussed here have been used extensively throughout this book, making it imperative that one develop a clear understanding of them at the very outset. The following section should provide a macro-level perspective of the synthesis process, and the background needed to invoke DC and execute the basic commands. Subsequent chapters discuss the steps in greater detail.

### 1.2.1 Design Analyzer and Design Compiler

Design Analyzer (referred to as DA for short) is the graphical front end of the Synopsys synthesis tool. Design Compiler or dc_shell is the command line interface for the same synthesis tool. In most cases, designers begin using the graphical front end, and once they are comfortable with the commands and the Design Compiler terminology, they prefer to use the command line (dc_shell) interface. At that stage, DA is generally used only to view schematics and their critical paths. The command line interface is identified by the following prompt:

```
dc_shell >
```

If the SYNOPSYS environment variable has been set to the synopsys root directory, then typing:

```
$SYNOPSYS/sparc/syn/bin/dc_shell
```

invokes DC and shows the dc_shell prompt. If not running on a sparc, use the corresponding architecture. A similar prompt can be seen from the Design Analyzer command window. The command window can be invoked from the Design Analyzer from the Setup -> Command Window pull-down menu.

```
design_analyzer >
```

When using the Design Analyzer, the command window helps the user understand the commands executed when using the menus in DA. DA, in turn, can be invoked by typing the following:

```
$SYNOPSYS/sparc/syn/bin/design_analyzer
```

## Startup Files

When DC is invoked, it reads the .synopsys_dc.setup file. The synopsys directory tree has a system wide .synopsys_dc.setup file. This file is located in $SYNOPSYS/admin/setup directory. In addition to this system wide file, the user can have a local .synopsys_dc.setup file in the current working directory or in the home directory. In general, the .synopsys_dc.setup file is used to specify certain commonly used variables like the target_library, link_library and search_path. The .synopsys_dc.setup file in the current working directory has the highest precedence, followed by the one in the user's home directory and finally, the system wide file. It is recommended that the user maintain a .synopsys_dc.setup file in the current working directory, since project or design specific variables can be specified without affecting other projects and designs. Shown below is a sample .synopsys_dc.setup file.

### Example 1.6 Sample Setup File

```
search_path= search_path+{".","./lib","./vhdl","./script}
target_library = {target.db}
link_library = {link.db}
symbol_library = {symbol.sdb}
```

DC follows the paths in the search_path variable from left to right. For example, if in a link_library file, link.db exists in the lib directory and in the vhdl directory, then the link.db file found in the lib directory is used. If the libraries are assigned correctly and the search_path indicates the location of these files, on invoking DA, the Setup -> Default pull-down menu should indicate the specified target, link and symbol libraries.

*Note: After invoking DA/DC, if any changes are made to the* .synopsys_dc.setup *file, be sure to include the* .synopsys_dc.setup *file using the* include *command at the dc_shell prompt (or the design_analyzer prompt in the DA command window) for these changes to take effect.*

```
include .synopsys_dc.setup
```

The target_library, link_library and search_path can be specified from DA, Setup -> Defaults menu. However, specifying them in .synopsys_dc.setup file saves one the trouble of specifying them each time DA is invoked. To verify the current value of any variable, use the list <variable_name> at the dc_shell prompt as shown below.

```
list target_library
```

## 1.2.2 Target Library, Link Library, and Symbol Library

Target library is the ASIC vendor library whose cells are used to generate a netlist for the design described in HDL during synthesis. The HDL code is *mapped* to cells from this library. The link library is used when a design is already in the form of a netlist or when the source HDL has cells instantiated from the technology library. The netlist is a design described using technology library cells. The link library which is specified by setting the link_library variable, indicates to DC, the library in which the descriptions of these cells are available. Similarly, the target_library variable and the symbol_library variable help specify the target and symbol libraries respectively.

Example 1.7 shows a simple Verilog netlist written out by DC. (Figure 1.11). This example should help to better understand the link library concept. The netlist has four instances (U5, U6, U7, U8) of the IVA library cell. If one wished to read the above netlist into DC and execute a command like report_timing, the link library declared by the link_library variable must have a description for the IVA library cell. If DC is unable to find a description for the IVA cell in the link library, the tool will be *unable to resolve the reference IVA*.

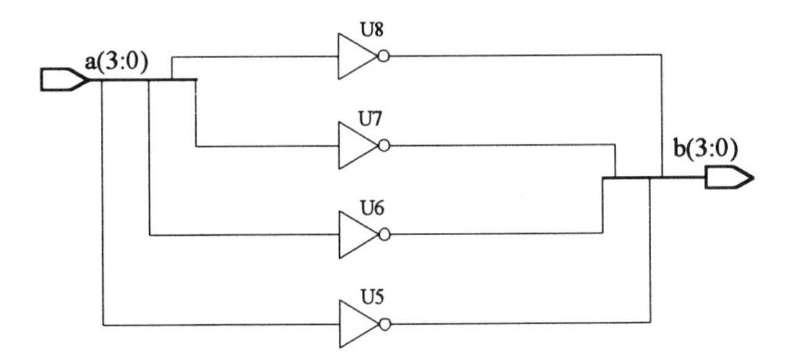


**Figure 1.11 Netlist Showing Instances**

### Example 1.7 Verilog Netlist Written by DC

```
module bus( a, b );
input  [3:0] a;
output [3:0] b;
   IVA U7 ( .A(a[2]), .Z(b[2]) );
   IVA U8 ( .A(a[3]), .Z(b[3]) );
   IVA U5 ( .A(a[0]), .Z(b[0]) );
   IVA U6 ( .A(a[1]), .Z(b[1]) );
endmodule
```

Symbol libraries are pictorial representations of library cells. If one wishes to view the schematic of a design in DA, then the tool requires a symbol library which contains actual graphic representations of all the library cells. The tool performs a one to one mapping of the cells in the netlist to cells in the symbol library when the user attempts to view the netlist representation. For example, on reading the above netlist into DC, followed by the link command, the tool looks for the IVA cell in the link library to *resolve the reference* IVA. Once it finds IVA in the **link_library**. the tool then looks for the IVA symbol in the symbol library. Viewing the schematic in DA can be done by double clicking on the design icon. However, the equivalent command executed is the **create_schematic** command. If it is unable to find cells in the symbol library, DA uses the generic symbol library (**generic.sdb**) to create schematics.

The technology and symbol libraries must exactly match in *case* for the cell names and their pin names. In other words, if the pin names of the IVA cell in the technology library do not match the pin names of the cell IVA in the symbol library, the tool will not be able to use the IVA symbol. In such a scenario, the tool uses symbols from the synopsys default generic library. The **compare_lib** command in DC, is a fast check mechanism to determine any differences between the symbol library and the technology library that might exist.

```
compare_lib <target_library>  <symbol_library>
```

The link mechanism can be forced to be case insensitive by setting the following **dc_shell** variable to false.

```
link_force_case = false
```

As shown in the sample **.synopsys_dc.setup** file in Example 1.6, the **target_library**, **link_library** and **symbol_library** are DC variables. The target and link libraries are of **.db** extension while **symbol_libraries** are of **.sdb** extension. Technology libraries are generated by the Synopsys Library Compiler from **.lib** files. These in turn are text files created by the ASIC vendor in Synopsys Library Compiler syntax. The ASIC vendor provides the user with **.db** and the **.sdb** files.

```
read_lib my_lib.lib
write_lib my_lib.db
```

Symbol libraries are created from .slib files.

```
read_lib my_lib.slib
write_lib my_lib.sdb
```

### 1.2.3 Cell names, Instance names, and References

In DC terminology, cell names and instance names are the same. References are either sub-designs or cells from a technology library. For example, if a design uses an IVA library cell as in Example 1.7, the tool provides it with an instance name (or cell name) such as U6. IVA is the reference and U1 is an instance of the IVA reference, which in turn is a library cell.

### Example 1.8 Sub-Design Instantiated in a Higher Level Block

**VHDL Code**

```
entity top is
port (a : in bit;
      b : out bit);
end top;
architecture test of top is
component sub1
        a : in std_logic ;
        z : out std_logic ;
end component;
begin
   U1 : sub1 port map (a => a, z=> b);
end test;
```

**Verilog Code**

```
module top( a, b );
input  [3:0] a;
output [3:0] b;
   sub1 U1 ( .A(a), .Z(b) );
endmodule
```

In Example 1.8, a sub design **sub1** is instantiated in a hierarchical block, **top**. When sub-designs are instantiated in higher level designs in a hierarchy, the name of the sub-design is the reference name (**sub1** in Example 1.8), while the instance name (or cell name) is the name assigned in the HDL code of the top level (U1 in this case) design during instantiation.

Two useful commands are report_reference and report_cell. These commands are helpful when one is required to find the references and instances inferred in a design after compile. For example, if a design comprises two library cells, each of which is used three times, report_cell command will list all the six instances and point to the corresponding reference, while the report_reference command will list just two references and the number of times each reference is used, as shown below. The distinction between cells, instances and references is extremely significant, particularly when writing dc_shell scripts. Shown below are the outputs of the report_cell and report_reference commands on the netlist discussed in Example 1.6. Notice that the report_reference shows just one reference, while the report_cell shows four instances or cells.

```
dc_shell> report_cell
****************************************
Report : cell
Design : test_bus
****************************************

Attributes:
    b - black box (unknown)
    h - hierarchical
    n - noncombinational
    r - removable
    u - contains unmapped logic

Cell        Reference     Library     Area   Attributes
-----------------------------------------------------------------
U5            IVA           lsi_10k      1.00
U6            IVA           lsi_10k      1.00
U7            IVA           lsi_10k      1.00
U8            IVA           lsi_10k      1.00
-----------------------------------------------------------------
Total 4 cells                         4.00
```

```
dc_shell> report_reference
****************************************
Report : reference
Design : test_bus
****************************************

Attributes:
    b - black box (unknown)
   bo - allows boundary optimization
    d - dont_touch
   mo - map_only
    h - hierarchical
    n - noncombinational
    r - removable
    s - synthetic operator
    u - contains unmapped logic

Reference  Library  Unit Area   Count   Total Area   Attributes
---------------------------------------------------------------------
IVA        lsi_10k  1.000000    4       4.000000
---------------------------------------------------------------------
Total 1 references  4.000000
```

### 1.2.4 VHDL Libraries in the Synthesis Environment

The VHDL language supports libraries. That is, frequently used functions and component declarations are stored in packages and these packages are analyzed into libraries. The packages are then called via the `use` clause in VHDL. A package must be analyzed prior to being used in a another design. The package can be a part of the VHDL code or a separate VHDL file. If the package is a separate file, it must be analyzed prior to being used in a design. In general, it is recommended that one maintain separate package files and declare them using the `use` clause when required in VHDL design files. Since Verilog does not have a configuration management mechanism like VHDL, this is not applicable to Verilog.

### Example 1.9 A Package, a VHDL Design, and a dc_shell Script

```
package my_pack is
type fsm_states is (state1, state2, state3, state4);
end my_pack;
// VHDL  file using my_pack
library States;
use States.my_pack.all;
```

Analyzing (using the analyze command as shown in Example 1.10) a design described in VHDL in DC, results in an intermediate format of files with .syn, .sim and .mra extensions. Example 1.9 shows a package my_pack, a VHDL design file that requires the my_pack package and the dc_shell script. The synthesis tool provides a mechanism by which the user can map a design library to a UNIX directory. Maintaining libraries is a good design practice as it simplifies the process of managing files.

### Example 1.10 dc_shell Script

```
define_design_lib States -path ./lib
analyze -format vhdl my_pack.vhd -lib States
```

Example 1.10 shows the dc_shell script that maps the States VHDL library to the UNIX directory lib in the current working directory. On executing the analyze command, the intermediate files are placed in the lib directory. This is extremely useful because it prevents the working directory from being cluttered with files. If the analyze command was used without the -lib States option, by default the intermediate files are written to the work library. The work library is mapped to the current working directory by default. To over-ride this default, one must use the define_design_lib command to map the work library to a particular unix directory. In general, it is recommended that one create a directory called work in the current working directory and map the work library to it. To verify the above steps, execute the report_design_lib command at the dc_shell prompt:

**dc_shell >** report_design_lib States

Shown below is the output on executing report_design_lib ieee.

```
****************************************
Report : hdl libraries
****************************************
Contents of current design libraries
```

```
ieee ($SYNOPSYS/sparc/packages packages/IEEE/lib)
   package:  n  STD_LOGIC_SIGNED
   package:  n  STD_LOGIC_UNSIGNED
   package:  n  std_logic_1164
   package:  n  std_logic_arith
   package:  n  std_logic_misc
```

VHDL Compiler does not support configuration declarations. Hence for synthesis one cannot have different entities in a design analyzed in different VHDL design libraries. Since packages are made visible by the use clause, they can be analyzed into different design libraries. To analyze design entities in different VHDL design libraries one must elaborate them individually and link their db files. Say for example you have a top level entity top which instantiates an entity leaf. Say, design entity leaf is analyzed into design library lib1 and top into the work library. Then the entity leaf in the library lib1 must be elaborated using the following command

```
elaborate leaf -library lib1
```

This will create a db file for the design leaf which can be saved and used when linking in the design top.

### 1.2.5 Synthesis, Optimization and Compile

Synthesis is the all encompassing, generic, (and now omnipresent in IC design) term for the process of achieving an optimal gate level netlist from HDL code. Therefore, synthesis includes both reading in the source HDL and optimization of the code. Optimization the other hand, is a step in the synthesis process which ensures the best possible combination of library cells which meet the functional, area and speed requirements of the target design. Compile is the process which executes the optimization step.

In DC, compile is the command which executes the optimization. After the source VHDL or Verilog has been read into DC, on executing the compile command, a netlist for the source HDL is generated. Before executing the compile step, the target library must be specified, if it is not already specified in the .synopsys_dc.setup file. Optimization constraints must be specified prior to compile. Compile has a number of options, including low, medium and high map efforts. The default option is a medium effort compile. In general, if one were merely running tests to check the logic inferred on compile, one should use the low map effort since it takes the least run time. The medium effort is recommended in most cases. The high map effort takes significantly longer compile run time.

### Example 1.11 dc_shell Script

```
read -format vhdl test.vhd
include constraints.scr
compile
write -f vhdl test -output test_netlist.vhd
```

Example 1.11 shows a simple dc_shell script to read in a design, compile, and write out a netlist of the design in VHDL. The file constraints.scr must contain timing and area constraints. Optimization constraints are discussed in greater detail in Chapter 4. For man pages on any of the DC commands, use the help command at the dc_shell prompt as shown below:

```
help report_reference
```

### 1.2.6 Synopsys_cache

DC maintains a synopsys_cache directory, where DW components which are inferred (or instantiated) from technology library cells are saved (or written out). This prevents the need to re-build these components, each time a compile step is executed. Reading and writing to this location is controlled by the cache_read and cache_write variables. Also, if the cache_read_info and cache_write_info variable are set to true, DC issues a message each time the cache is written to, or read from.

## 1.3 Classic Scenarios

In this section, we describe a few classic scenarios faced by designers when using DC. This section is especially useful for designers new to DC.

**Scenario 1**

You are linking a design and DC issues one of the following warnings:

```
Warning :   Unable to resolve reference xxx in yyy.  (LINK-5)
Warning :   Design test  has 3 unresolved references.(UID-341)
```

**Solution**

These warnings could be because of one of the following reasons:

1. The search_path variable is incorrectly specified. Check the search_path variable to verify that the UNIX paths to the target and link library have been specified correctly. At the prompt use the following command:

   ```
   list search_path
   ```

2. The reference xxx is a subdesign that is instantiated in yyy with instance name u1. xxx does not have a db file which exists in one of the directories in the search_path Read in the source HDL for xxx before reading in the HDL for yyy. If a db file already exists for reference xxx, modify the search_path variable to include the path to the db file.

**Scenario 2**

DC issues one of these warnings when executing a report.

```
Warning :   Can't resolve reference LD1 for cell u1  (UID-233)
Warning :   Can't find symbol ..."
```

**Solution**

This is similar to scenario 1. Once again, check to see if the search_path does include the path to the link library and the target library. If it does include the path to the link library and target library, check to see if an instantiated component in the HDL does exist in the link_library, symbol_library and target_library. Notice that the warning points to the reference LD1. To verify that the LD1 library cell exists in the technology library libA use the following command:

```
find(cell, libA/LD1)
```

If the find command results in an error message, use the list -libraries command to find the exact name of the library. The library name could be libA.db instead of libA. Every instantiated component must be referenced to a sub-design or a library cell. If it is a library cell, the target technology library or the link library must have a description for it. If it is a sub design, the source HDL for the sub design must be read into DC, prior to reading in this file. If a db file exists for this sub-design, the search path must include a path to this db file.

**Scenario 3**

DC issues the following error on reading in the source VHDL into DC. The library declaration in the VHDL file is as shown below.

```
library test ;
use test.pack.all;
Error: The library 'test' is not mapped to a directory (LBR-6)
```

**Solution**

This error occurs when the VHDL design library test is not mapped to a valid unix directory. Use the following command at the dc_shell prompt.

```
define_design_lib test -path <the unix path to library files>
```

**Scenario 4**

DC issues the following warning message when reading in a VHDL file.

Warning: The library 'test' is mapped to the directory 'lib' which is not writable. The strict VHDL analyzer will not be able to be invoked. (HDL-213).

**Solution**

This error occurs when one has executed the define_design_lib test -path ./lib command but there does not exist a directory lib in the specified location, or it exists but the user does not have write permission in that directory.

**Scenario 5**

During compile, DC issues the following warning:

Warning: The cache_write directory xxx is not writable. (SYNOPT-11)

**Solution**

Reading and writing to the cache is controlled by the cache_read and cache_write variables. Ensure that these variables point to your home directory or to any shared cache that might exist for your design team.

**Scenario 6**

When reading in the design database (db file), DC issues the following error.

Error: db file is corrupted. (EXPT-18)
What could be the cause?

**Solution**

It likely that you are using different versions of Synopsys tools. In other words, the db file was generated by a more recent version of DC while you are trying to read the db file now into a previous version of DC. In short, db files are backward compatible but not forward compatible. To check the version of DC, use the following command:

list product_version

**Scenario 7**

In DA, the schematic shows mere boxes instead of the actual symbols for gates.

**Solution**

Ensure that the symbol library (.sdb file extension) for the technology library is available and specified by the symbol_library variable. Also, verify that the search_path variable in your .synopsys_dc.setup file includes the path to where the symbol library file is located. After doing the above, execute the following steps to verify:

```
read -f db <your_sdb_file>
create_schematic /* Or select View -> Recreate in DA*/
```

If DA still shows boxes instead of symbols, it is likely that the symbol library and the technology library do not match in case for the pin names or cell names. Execute the compare_lib command to verify the same as shown in section1.2.3.

**Scenario 8**

When typing $SYNOPSYS/sparc/syn/bin/design_analyzer to invoke DA, you get the following message:

```
Error : DC not enabled. (DCSH-1).
```

**Solution**

By default, DC expects the key file to be located in the $SYNOPSYS/admin/license directory. The SYNOPSYS_KEY_FILE environment variable can be used to specify a different location for the key file. Verify your environment variable SYNOPSYS_KEY_FILE. Ensure that it points to $SYNOPSYS/admin/license/key or the appropriate key file. If this is indeed the case, the key file may have a "typo", or else you do not have the appropriate keys.

## Recommended Further Readings

1. Design Compiler Family Reference Manual
2. Designware User Guide

CHAPTER 2

# *VHDL/Verilog Coding for Synthesis*

This chapter provides several examples of coding for synthesis in both VHDL and Verilog. First, general HDL coding issues are discussed, followed by a brief comparison of the two commonly used languages VHDL and Verilog. Then, coding for finite state machines has been discussed in detail with several examples followed by specific tips for FSM coding. Examples in both VHDL and Verilog have been provided to infer a multi-bit register, barrel shifter, incrementor etc. Finally, a few classic scenarios have been discussed.

## 2.1 General HDL Coding Issues

In this section, we discuss some basic issues related to HDL coding for synthesis such as VHDL types, unwanted latches, variables and signals, and priority encoding. For a certain desired functionality, it is often possible to code HDL in a number of different ways. However, there are several guidelines that one can follow to develop a consistent coding style for synthesis.

### 2.1.1 VHDL Types

It is recommended that std_logic types be used for port declarations in the entity. This convention avoids the need for conversion functions when integrating different levels of synthesized netlist back into the design hierarchy. The *std_logic* type is declared in the IEEE std_logic_1164 package. Some of the examples in this chapter use the type bit for the sake of simplicity and easy understanding.

The type buffer can be used when an output must be used internally. The use of mode buffer is not recommended for synthesis. This is because ports of mode buffer can only be associated with ports of mode buffer and gate level VHDL simulation models from ASIC vendors never use the mode buffer. Once declared as a buffer, all references to the particular output port must be declared as buffer throughout the hierarchy. Hence, there is a potential for a port mode type mismatch when using mode buffer. This is often overlooked and one can run into problems during integration of different blocks. For the sake of consistency, it is recommended that one avoid the use of the type buffer.

### Example 2.1 VHDL Coding Style to Avoid Buffer Data Types

```
entity buf is
port (a : in std_logic ;
      b : in std_logic ;
      c : out std_logic );
end buf ;
architecture test of buf is
signal c_int : std_logic ;
-- without c_int signal c will have to be declared as buffer
begin
process
begin
    c_int <= a + b + c_int ;
end process;
    c <= c_int ;
end buf ;
```

Example 2.1 shows an effective way to avoid the use of buffer types using internal signal declarations.

## 2.1.2 Signals and Variables

During simulation, variables are updated immediately unlike signals which require a delta time before being updated. Variables tend to simulate faster than signals but could mask glitches that might affect the functionality of the design. Further, variables tend to generate unexpected results during simulation.

All signals that are being read in a process (on the right side of a VHDL assignment) when not declared in the process sensitivity list could cause a (simulation/synthesis mis-match) mis-match between RTL simulation and the gate-level simulation of the synthesized netlist. This causes DC to issue a warning that signals which are being read are not in the sensitivity list, or in other words, will not trigger the process.

### 2.1.3 Priority Encoding Structure

Coding with if statements causes priority encoding logic to be inferred. In other words, DC assumes that the first if condition has a greater priority than the second and so on. To avoid this one must use case statement. Refer to Scenario 3 in section 2.1.

### 2.1.4 Unwanted latches

Ensure that all signals are initialized. Further, when using case statements or nested if statements, ensure that they are fully defined. A full specification will prevent latches from being inferred.

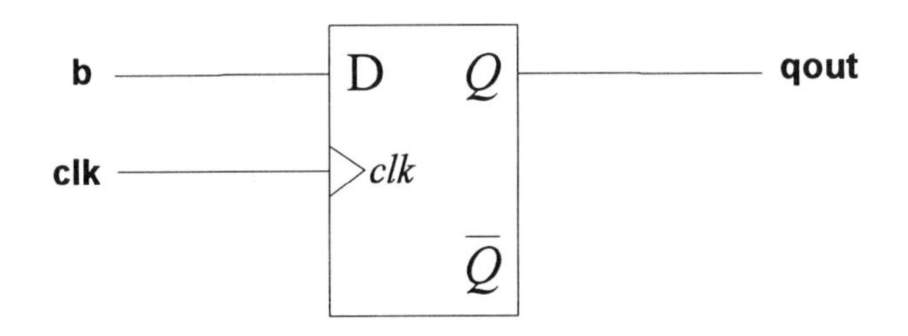


**Figure 2.1 Latch**

#### Example 2.2 Code for Latch

**VHDL Code**

```
library ieee;
use ieee.std_logic_1164.all;

entity latch is
  port ( clk       : in std_logic;
         b         : in std_logic_vector (2 downto 0 );
         qout      : out std_logic_vector (2 downto 0 ) );
end latch;
architecture behavior of latch is
begin
process (clk, b)
```

```
begin
        if (clk = '1') then
            qout <= b;
        end if ;
end process;
end behavior;
```

**Verilog Code**

```
module latch(clk,qout,b);
input clk, b;
output qout;
reg q;
always @ (clk or qout)
      if (clk == 1)
qout = b;
endmodule
```

Consider the Example 2.2 shown above. Notice that the expected result when clk is not equal to '1' is not specified. DC interprets this to mean that when clk=1 condition is not satisfied, retain the previous value of q. Hence, latches are inferred. After reading in the HDL, one does not have to compile the design to realize that unwanted latches have been inferred. The dc_shell output provides information on memory devices inferred as shown below:

```
dc_shell > read -f vhdl test.vhd
Loading vhdl file '/home/test.vhd'
/home/test.vhd:
Statistics for inferred devices in process at line 15 in file /home/test.vhd
===========================================================
|          | Three  | Memory |           |
|Variable | State   | Devices | Width | Conditionally Driven (Line #) 7        |
===========================================================
| q       | No     | Latch    | 3   |         Yes (17)                         |
===========================================================
Current design is now 'test_width4'
```

### 2.1.5 Synchronous and Asynchronous Resets

Designs often require a synchronous or an asynchronous reset signal to be inferred. In this section we show how asynchronous (Example 2.3) and synchronous (Example 2.4) resets can be described. The only difference between inferring synchronous and asynchronous resets is the declaration of the if statement. Notice that the if statement appears before the clock statement for asynchronous reset, and after the clock for synchronous resets.

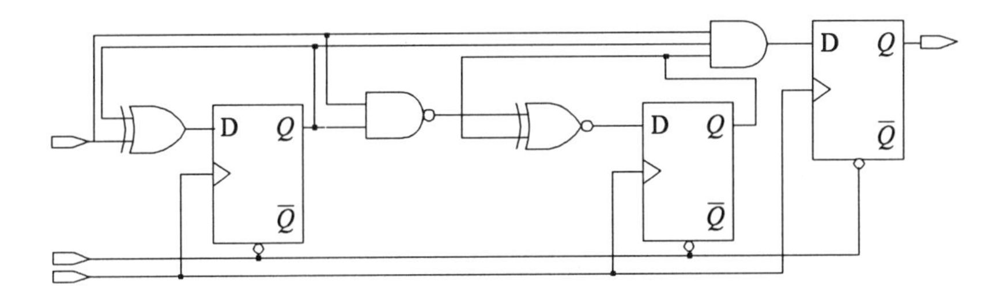


**Figure 2.2 Design With Asynchronous Reset**

## Example 2.3 HDL Code Showing Asynchronous Reset

```
entity test is
port (X, clock, RST : in bit;
            Z : out bit);
end test;
architecture trial of test is
begin
process
begin
        if RST = '0' then
                ST <= S0 ;
                 Z <= '0';
        elsif clock' event and clock = '1' then
        if X='0'
                then Z <= '0';
        else
        case ST is
-- case statement same as in example 2.1
        end case ;
      end if;
end process;
end trial ;
```

**Verilog Code**

```
module test (X, clock, rst, Z);
input X, clock, rst;
output Z;
reg Z ;
// state can take one of these values
parameter S0 = 2'b00, S1 = 2'b01,S2 = 2'b10,S3 = 2'b11;
reg [1:0]  ST;
/* Sequential logic of FSM */
always @(posedge clock or negedge rst)
begin
  if (!rst) begin
       Z <= 1'b0;
       ST <= S0;
  end
  else begin  // posedge clock is assumed
       Z <= 1'b0;
       if (X == 1'b0) begin
          Z <= 1'b0;
          ST <= ST;
        end
      else begin
          case (ST)
          // case statement same as in example 2.1
          endcase
      end
  end
end
endmodule
```

Example 2.3 shows how an asynchronous reset can be incorporated into the design. (Figure 2.2). Example 2.3 is similar to Example 2.4, except that an additional input for reset (RST) is declared in the entity and the reset condition is added before the clock' event and clock = 1; statement, implying that the reset signal is asynchronous.

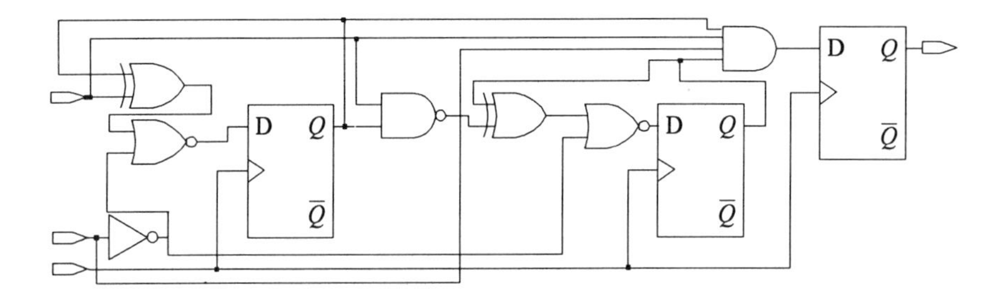


**Figure 2.3 Design With Synchronous Reset**

## Example 2.4 HDL Code Showing Synchronous Reset

**VHDL Code**

```
entity test is
port (X, clock, RST : in bit;
        Z : out bit);
end test;
architecture trial of test is
begin
process
begin
wait until clock' event and clock = '1';
if RST = '0' then
       ST <= S0 ;
        Z <= '0';
elsif
       X='0'  then
       Z <= '0';
else
     case ST is
-- case statement same as in example 2.1
     end case;
end if;
end process;
end trial ;
```

**Verilog Code**

```
module test (X, clock, rst, Z);
input X, clock, rst;
output Z;
reg Z ;
// state can take one of these values
parameter S0 = 2'b00, S1 = 2'b01,S2 = 2'b10,S3 = 2'b11;
reg [1:0]  ST;
/* Sequential logic of FSM */
always @(posedge clock)
begin
  if (!rst) begin
       Z = 1'b0;
      ST = S0;
  end
  else begin
       Z = 1'b0;
      if (X == 1'b0) begin
         Z = 1'b0;
         ST = ST;
      end
      else begin
          case (ST)
           // case statement same as in example 2.1
          endcase
       end
  end
end
endmodule
```

## 2.2 VHDL vs. Verilog: The Language Issue

Which HDL should one use? Does synthesis favor one over the other? In this section, we provide practical comparisons between the two leading HDLs, with coding for logic synthesis in perspective. Both VHDL and Verilog HDL are the leading Hardware Description Languages and will continue to co-exist. Since our first edition, Verilog has become an IEEE standard, thereby eliminating one more advantage which VHDL had over Verilog. VHDL development was sponsored by the Department of Defense(DOD) and VHDL models are a required deliverable for all DOD sponsored digital designs. The intent of this section is to discuss the most obvious differences in coding that one has to be aware of when coding in either VHDL or Verilog, but not to favor one over the other.

## Case Sensitivity

VHDL is a case insensitive language while Verilog is a case sensitive language.

## Case Statements

By definition, the different branches in a case statement in VHDL are mutually exclusive. This implies that when synthesizing a **case** statement in VHDL, there is no priority logic implied. In contrast, the different choices of a **case** statement in Verilog are not by definition mutually exclusive. Hence, you can use the Verilog case statement to synthesize priority logic. In order to explicitly specify that the different choices in the **case** statement are mutually exclusive one can use the Synopsys pragma **parallel_case** in the Verilog code. A common application of this is when designing a state machine using Verilog. Since the state machine can't be in more than one state at a given time, the case statement used to decode the states of the state machine must use the **parallel_case** pragma for synthesis.

VHDL, by definition, requires that all the different values of an expression in a **case** statement be covered. However, this is not required in Verilog. If the user knows that all possible values of the case expression have been covered, one should use the Synopsys synthesis **full_case** pragma. This prevents latches from being inferred due to not assigning values to a reg/signal under all possible conditions.

## Sign Interpretation of Data Types

In VHDL, based on the data type used and the functions defined for this datatype, the operations associated with the data type are interpreted as signed or unsigned. For example, the data type **std_logic_vector** defined by IEEE is not defined as either signed magnitude or unsigned magnitude. Depending on how you wish to interpret them you might use the corresponding functions. Synopsys provides two packages **std_logic_unsigned** and **std_logic_signed** which have overloaded functions for the data type **std_logic_vector**. The Verilog language interprets a register data type as an unsigned value and an integer type as a signed value. Hence, when a negative value is assigned to a register and later used in an expression, it is interpreted as a positive number which is the 2's complement of the negative value.

## Example 2.5 Verilog Code Using Negative Numbers

```
reg [3:0] A;

always
begin
...
  A = -4'd7;
  B = A + C;
end
```

In Example 2.5, A will be assigned the value 1001 which is the 2's complement of -7, but interpreted as +9 in the expression A + C. One way to get around this is shown below.

```
if (A[3]) /* then a negative number */
  B = C - (~(A) + 1'b1);
else
  B = A + C;
```

## I/O Modes

VHDL permits ports of mode in, out, inout, and buffer. Verilog supports ports of mode input, output, and inout. VHDL does not allow reading ports of mode out internally in the design. Ports of mode output can be read in Verilog. In Verilog, this translates to creation of a temporary signal with the same name as the output port and this temporary signal net is read internal to the code. In VHDL for such a design scenario it is recommended that you explicitly create a temporary internal signal which is associated with the port of mode out. This signal can be read as well as assigned from within the code. A typical scenario for this is a counter where you need to read the current count of the counter which is an output port. Using ports of mode buffer in VHDL, allows you to read ports as well as write to it. They are not inout ports and can't be driven externally. For synthesis, use of ports of mode buffer is not recommended as discussed in section 2.1.

## Component Declarations

In VHDL, a component declaration is required before instantiating a component. This defines the template of the component being instantiated and includes information such as, port mode or direction, port names and the data type of the ports. These templates or component declarations are bound to sub-design entities using configuration declaration statements. Hence, when reading in a VHDL netlist you have the necessary information about the interface of the component and the respective directions and datatypes of the port thereby allowing it to be checked for consistency. Here is an example of a VHDL netlist instantiating a buffer from a technology library.

## Example 2.6 Component Instantiation

### VHDL Code

```
library IEEE;
use IEEE.std_logic_1164.all;

entity top is
  port (A: in std_logic;
        B: out std_logic);
end top;

architecture struct of top is

-- COMPONENT DECLARATION

component BUFA
  port (A: in std_logic; B: out std_logic);
end component;

begin

I1: BUFA port map( A => A, B => B);

end struct;
```

### Verilog Code

```
module top (A, B);
input A;
output B;
  BUFA I1(A, B);  // Component Instantiation
endmodule
```

Example 2.6 shows a Verilog description which instantiates the same buffer cell from the technology library. In Verilog, there is no concept of component declarations. The direction and datatypes of the ports of the component being instantiated are derived from the *module* to which it is linked.

### Multiple Drivers

When there are multiple drivers on a net, a resolution function is needed to resolve the final value on the net. In VHDL, one can create resolution functions and associate it with a datatype or a signal object. std_ulogic is an example of an unresolved datatype, while std_logic is an example of a resolved datatype. The resolution function is used to figure out the final value on a signal when there are more than one driver on it. In Verilog, one can't have user defined resolution functions. Instead one can use net types with built in resolution functions (pre-defined in Verilog) such as wand, wor, and so on.

## 2.3 Finite State Machines

The design issues in the coding of state machines like state encoding, registered outputs, synchronous resets, asynchronous resets and *fail-safe* behavior of state machines are crucial for effective synthesis of state machines. In this section, the design and synthesis of clocked synchronous state machines using DC are discussed along with several examples. The advantages and disadvantages of the different techniques are also discussed. Finally some tips and limitations of state machine synthesis are provided.

A finite state machine (FSM) consists of a current state (P) and a next state (N), inputs (I) and outputs (O). State Machines can be classified as *Mealy* or *Moore* machines depending on how the outputs are generated.

A sequential state machine whose outputs depend on both the current state and the inputs is called a Mealy machine, as shown in Figure 2.4. In other words, the functionality can be expressed as,

Next state (N) = function [current state (P), Inputs (I)]
Outputs (O) = function [current state (P), Inputs (I)]

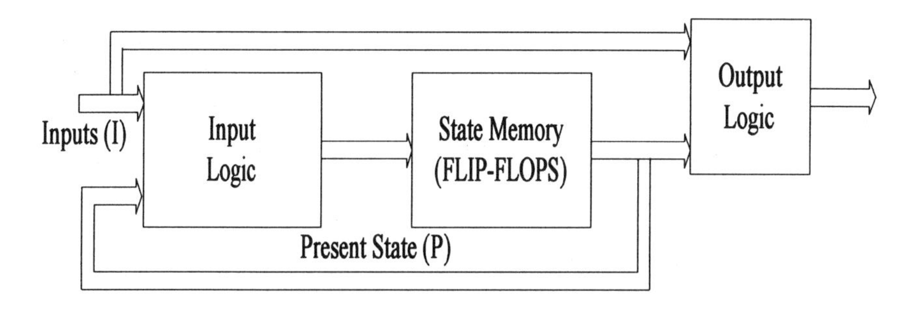


**Figure 2.4 Mealy Machine**

A Moore machine is one in which the outputs are a function of only the current state and independent of the inputs (Figure 2.5). In other words, the functionality can be expressed as,

Next state (N) = function [current state (P), Inputs (I)]

Outputs (O) = function [current state (P)]

The next state logic and the output logic are purely combinational while the present state consists of sequential memory elements (flip-flops). Each active clock transition causes a change of state from the present state to the next state.

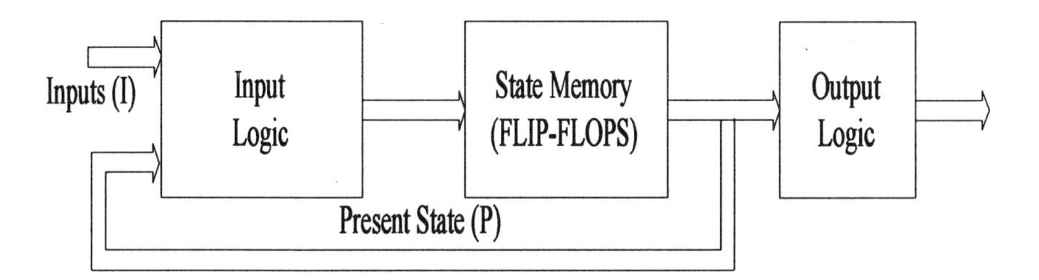

**Figure 2.5 Moore Machine**

### State Encoding

The concepts of current state and next state are vital to any state machine. Flip flops in a state machine serve as memory elements keeping track of the current state. Each possible state of the state machine can be assigned a unique binary code. This is called state encoding. At any given instant, the current state of the state machine is determined by the binary values in the flip flops and their corresponding encoding. Thus n flip flops will encode a maximum of $2^n$ states. Alternatively, one can assign one flip flop to each state. Thus n flip-flops will represent n states. This is called the *One-hot* method of encoding. Since the state machine can only be in one state at a given time, the outputs of only one of the flip flops is true and hence the name One-hot. The use of one flip flop for each state could result in greater silicon area. The advantage of this method lies in that no combinational logic is required to decode the values of the current state in the state flip flops, since each state has only one flip flop. This makes the one-hot state machine the fastest implementation.

## 2.3.1 VHDL/Verilog Coding for FSMs

In this section, we describe a Mealy machine and discuss two possible ways (Example 2.7 and Example 2.8) of coding the same in both VHDL and Verilog. Consider a Mealy machine with one input (X) and one output (Z). When X = 0, the current state of the state machine remains unchanged and output Z remains at 0. When X =1, the state machine makes a transition from one state to the next binary state, that is, 00 -> 01 -> 10 -> 11 -> 00 and so on. The output Z is equal to 1, only when the state is 11 and the input X is equal to 1, else Z is equal to 0 as shown in the state transition Table2.1 and state transition diagram in Figure 2-3.

**Table2.1 State Transition Table**

| Input (X) | PRESENT_STATE | NEXT_STATE | Output (Z) |
|---|---|---|---|
| 0 | $Q_i$ | $Q_i$ | 0 |
| 1 | $Q_i$ | $Q_{i+1}$ | 1 (when $Q_3$) else 0 |

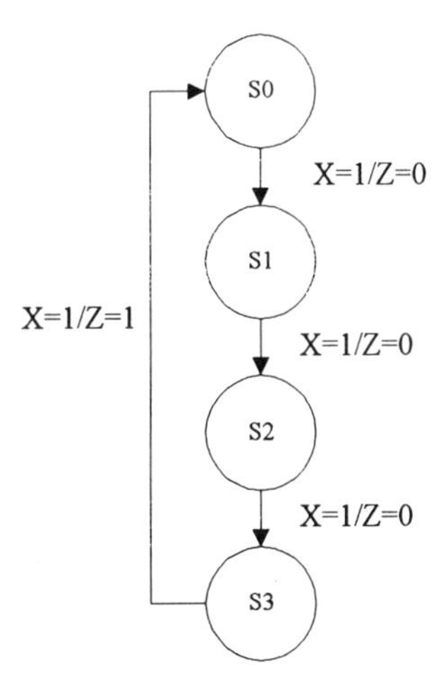


**Figure 2.6 State Transition Diagram**

## Example 2.7 HDL Description of State Machine

### VHDL Code

```
package states is
type state is (S0, S1, S2, S3); -- state can take one of these values.
end states;
use work.states.all;
entity test is
port (X, clock : in bit;
        Z : out bit);
end test;
architecture trial of test is
signal ST : state;
begin
  process
  begin
      wait until clock' event and clock = '1';
      if  X='0' then Z = 0;
     else
          case ST is
```

```
            when S0 => ST <= S1 ; Z <= '0';
            when S1 => ST <= S2 ;Z <= '0';
            when S2 => ST <= S3;Z <= '0';
            when S3 => ST <= S0 ;Z <= '1';
          end case;
        end if;
  end process;
end trial;
```

**Verilog Code**

```
module fsm1 (X, clock, Z);
input X, clock;
output Z;
reg Z ;
// state can take one of these values
parameter S0 = 2'b00, S1 = 2'b01,S2 = 2'b10,S3 = 2'b11;
reg [1:0]  ST;
/* Sequential logic of FSM */
always @(posedge clock)
begin
  Z = 1'b0;
  if (X == 1'b0) begin
       Z = 1'b0;
       ST = ST;
       end
   else
       case (ST)
         S0: ST = S1;
         S1: ST = S2;
         S2: ST = S3;
         S3: begin
              ST = S0;
              Z = 1'b1;
         end
       endcase
end
endmodule
```

Example 2.7 shows the state machine using a **case** statement and a **wait** statement. In the event of the input X being 0, the output Z always remains 0 and no state transition occurs. DC interprets that the state remains unchanged, if not mentioned. In other words, DC will maintain the current state.

## Example 2.8 Recommended Coding Style for State Machines

### VHDL Code

```
package states is
  type state is (S0, S1, S2, S3); -- state can take one of these values.
end states;
use work.states.all;
entity test is
port (X, clock : in bit;
        Z : out bit);
end test;
architecture trial of test is
signal CURRENT_STATE, NEXT_STATE: state;
attribute STATE_VECTOR : STRING;
attribute STATE_VECTOR of trial : architecture is "CURRENT_STATE";

begin
  COMB : process (CURRENT_STATE, X)
  begin
    case CURRENT_STATE is
    when S0 =>
    if X = '0' then
      Z <= '0';
      NEXT_STATE <= S0 ;
    else
      Z <= '0';
     NEXT_STATE <= S1;
    end if;
    when S1 =>
    if X = '0' then
      Z <= '0';
      NEXT_STATE <= S1 ;
    else
      Z <= '0';
```

```
        NEXT_STATE <= S2;
      end if;
      when S2 =>
      if  X = '0' then
        Z <= '0';
        NEXT_STATE <= S2 ;
      else
        Z <= '0';
        NEXT_STATE <= S3;
      end if;
      when S3 =>
      if  X = '0' then
        Z <= '0';
        NEXT_STATE <= S3 ;
      else
        Z <= '1';
        NEXT_STATE <= S0;
      end if;
      end case;
  end process;
SYNCH : process

begin
  wait until clock' event and clock='1';
  CURRENT_STATE <= NEXT_STATE;
end process;
end trial;
```

**Verilog Code**

```
module fsm (X, clock, Z);
input X, clock;
output Z;
reg Z ;
parameter [1:0]
S0 = 2'b00,
S1 = 2'b01,
S2 = 2'b10,
S3 = 2'b11;
```

```
reg [1:0]  CURRENT_STATE, NEXT_STATE;
//synopsys state_vector CURRENT_STATE
always @(posedge clock)

begin
     CURRENT_STATE = NEXT_STATE ;
end
always @(CURRENT_STATE or X)

begin
  case (CURRENT_STATE)
 S0: if (X == 1'b0) begin
      Z = 1'b0;
      NEXT_STATE = CURRENT_STATE;
      end
      else begin
      Z = 1'b0;
      NEXT_STATE = S1;
      end
 S1: if (X == 1'b0) begin
      Z = 1'b0;
      NEXT_STATE = CURRENT_STATE;
      end
      else begin
      Z = 1'b0;
      NEXT_STATE = S2;
      end
 S2: if (X == 1'b0) begin
      Z = 1'b0;
      NEXT_STATE = CURRENT_STATE;
      end
      else begin
      Z = 1'b0;
      NEXT_STATE = S3;
      end
 S3: if (X == 1'b0) begin
      Z = 1'b0;
      NEXT_STATE = CURRENT_STATE;
      end
```

```
        else begin
        Z = 1'b1;
        NEXT_STATE = S0;
        end
endcase
end
endmodule
```

Example 2.8 shows another approach to coding the same state machine. This form of coding tells DC that the design is a state machine without having to set the state vectors after reading in the design. This is possible by using the **state_vector** attribute. The **state_vector** attribute on the architecture is assigned a value which is the name of the state signal. The design has been divided into two separate processes. The first process COMB, describes the combinational part of the design, while the second process SYNCH, describes the synchronous part of the design.

### Registered Outputs

Outputs when generated by combinational logic could result in glitches. To avoid glitches in the outputs, designers infer registered outputs.

*Note: In Example 2.7 the output Z is a registered output while in Example 2.8, the output Z is not registered.*

One of the advantages of the approach shown in Example 2.8 is that it separates the synchronous part of the design from the combinational parts. This style of coding provides the designer complete control over the combinational and sequential parts of the design, making the debug process less complicated. Hence, the recommended FSM coding style for synthesis is to use two separate processes, one for the combinational and the other for the sequential parts of the design.

## 2.3.2 Enumerated Types and Enum_encoding

Another approach to coding FSMs in VHDL is by the use of *enumerated types* with the use of the enum_encoding attribute. In this approach, one must declare the list of all possible values of the type state (say, S0, S1, S2, and S3 as in Example 2.7). The values (S0, S1, S2, S3) can be either an identifier (sequence of letters, underscores and numbers) or a character literal ('A', 'B'). The VHDL Compiler does default encoding of the enumerated literals. That is, by default, the enumeration values are encoded into bit_vectors, the first enumerated literal being assigned 0, the next 1 and so on depending on the number of values. Minimum number of bits are used in the encoding. For example, to encode four states two bits are used. To override the default enumeration encodings, the enum_encoding attribute can be used. The enum_encoding attribute must immediately follow the type declaration and must be a string containing a series of vectors, one for each enumerated literal in the type declaration. Again, the first vector corresponds to the encoding for the first enumeration literal and so on.

## Example 2.9 Enumerated Types Using the enum_encoding Attribute

### VHDL Code

```
entity test is
port (X, clock : in bit;
      Z : out bit);
end test;
architecture trial of test is
type state is (S0, S1, S2, S3);
attribute ENUM_ENCODING : string;
attribute ENUM_ENCODING of state: type is "00 01 10 11";
signal CURRENT_STATE, NEXT_STATE: state;
signal  Z_int : bit;
begin
-- COMB and SYNCH processes same as in example 2.2
end trial;
```

### Verilog Code

```
module test (X, clock, Z);
input X, clock;
output Z;
reg Z ;
parameter [1:0] // synopsys enum states
S0 = 2'b00, S1 = 2'b01, S2 = 2'b10, S3 = 2'b11;
reg [1:0]  /* synopsys enum states */ CURRENT_STATE, NEXT_STATE;
// combinational and sequential blocks same as in example 2.2
endmodule
```

Example 2.9 shows the use of *enumerated types* with the use of the enum_encoding attribute. The declaration, type state is (S0, S1, S2, S3), defines the list of all possible values of the type state. By default, the enumeration values are encoded into bit_vectors, the first enumerated literal S0 being assigned 0, the S1 being assigned a 1 and so on. By using the enum_encoding attribute the encoding of the different states is declared in the code. This approach to coding FSMs is the recommended approach. The combinational and sequential parts are in separate processes. Further, by the use of enum_encoding, one has control over the states and their encodings.

### 2.3.3 One-hot Encoding

The fastest FSM implementation uses the one-hot method of encoding. In DC, the user will have to declare the FSM encoding style using the set_fsm_encoding_style command.

**Example 2.10 State Machine Using the state_vector and enum_encoding Attributes**

**VHDL Code**

```
package states is
  type state is (S0, S1, S2, S3); -- state can take one of these values.
  attribute ENUM_ENCODING : STRING;
  attribute ENUM_ENCODING of state : type is "0001 0010 0100 1000";
end states;
library work;
use work.states.all ;
entity test is
port (X, clock : in bit;
       Z : out bit);
end test;
architecture trial of test is
   attribute STATE_VECTOR : string;
   signal ST : state;
   attribute STATE_VECTOR of trial : architecture is "ST";
begin
-- COMB and SYNCH processes same as in example 2.2
end trial;
```

**Verilog Code**

```
module fsm (X, clock, Z);
input X, clock;
output Z;
reg Z ;
parameter [3:0] // synopsys enum states
```

```
    S0 = 4'b0001, S1 = 4'b0010, S2 = 4'b0100, S3 = 4'b1000;
reg [3:0] /* synopsys enum states */ CURRENT_STATE, NEXT_STATE;
//synopsys state_vector CURRENT_STATE
// combinational and sequential blocks same as in example 2.2
endmodule
```

Example 2.10 shows the same state machine described in Example 2.7 using both the `state_vector` attribute and the `enum_encoding` attribute. `enum_encoding` has been used such that the first flip-flop, when on, implies the state S0, the second flip-flop implies state S1, and so on. Note that the coding style uses the state vector attribute as in Example 2.8 and does not require the declaration of state vectors. In general, the one-hot encoding style involves the use of one flip-flop for each state, the current state being determined by the flip-flop which is on.

## Tips for FSM Coding

- All ports of the initial design must be either input or output ports. Inout ports cannot be represented in a state table.
- Separate state machines into separate designs. FSMs must be *extracted* (extraction, executed by the `extract` DC command, helps to write out the design from DC in a state table format) before DC can write out the FSM in the form of a state table. In general, designs with multiple state machines, where the output is determined by the state of several state machines, cannot be extracted.
- All state flip flops must be driven by the same clock. Thus, any clock tree instantiation or clock tree balancing should be done after FSM extraction.
- Use of the `others` clause in the `case` statement. Consider the state machine described in Example 2.8 without the state S3. In other words, the type declaration has only three possible values -- SO, S1 and S2; and all cases of "when S3" are not specified in the case statements in Example 2.8. This code is very much a legal VHDL description. However, the logic inferred by the VHDL Compiler will not ensure *fail-safe* behavior. In other words, should the state ever become S3 or the equivalent encoding "11" the FSM will not necessarily transition to a particular known state. The compiler will optimize the design, without generating logic for possible non-declared states. Hence, it is essential in cases where *fail-safe* behavior is required, that the `others` clause be declared in the case statement. When the `others` clause is declared, the compiler will generate the required logic to reset the state machine to a known state when an unknown state is reached.

## Example 2.11 FSM Showing Fail/Safe Behavior

### VHDL Code

```
package states is
  type state is (S0, S1, S2); -- state can take one of these values.
end states;
use work.states.all;
entity test is
port (X, clock : in bit;
      Z : out bit);
end test;
architecture trial of test is
  signal ST : state;
begin
  process
  begin
    wait until clock' event and clock = '1';
    if X='0' then Z = 0;
    else
    case ST is
    when S0 =>
      ST <= S1 ; Z <= '0';
    when S1 =>
      ST <= S2 ;Z <= '0';
    when S2 =>
      ST <= S0;Z <= '0';
    when others => ST <= S0 ;
    end case;
    end if;
  end process;
end trial;
```

### Verilog Code

```
module fsm (X, clock, Z);
input X, clock;
output Z;
reg Z ;
// state can take one of these values
parameter S0 = 2'b00, S1 = 2'b01,S2 = 2'b10;
```

```
reg [1:0]  ST;
/* Sequential logic of FSM */
always @(posedge clock)
begin
  Z = 1'b0;
  if (X == 1'b0) begin
     Z = 1'b0;
     ST = ST;
  end
  else
     case (ST)
      S0: ST = S1;
      S1: ST = S2;
      S2: ST = S0;
      default: ST = S0;
     endcase
end
endmodule
```

Example 2.11 describes an FSM which shows fail-safe behavior. The others clause ensures that if the FSM were to go to the state defined by encoding "11", then it is reset to the state S0.

## 2.4 HDL Coding Examples

This section shows several simple HDL coding examples of commonly used building blocks such as RAMs, ROMs, barrel shifters and incrementors. Examples are provided in both VHDL and Verilog.

### Memories

In general, DC is not used to synthesize memories except for small scratch pad memories. Instead, they are instantiated as hard macros (black boxes) during synthesis and incorporated into the design during layout. However, when RAMs/ROMs are used in designs, they require HDL descriptions for simulation. Similarly, for static timing analysis in Synopsys, you need to create a model of the memory using Synopsys Library Compiler syntax without any function attribute. In this section, we describe an example of a RAM model. Then, we show a sample Verilog code for a ROM for simulation purposes. This example shows the use of the synopsys pragmas that *turn off* synthesis of the ROM code.

## Example 2.12 HDL Description of a RAM

### VHDL Code

```
-- One Port Synchronous RAM with Read/Write Enable

library IEEE;
use IEEE.std_logic_1164.all;
use IEEE.std_logic_unsigned.all;

entity ram_vhdl is
  generic (width: natural := 8;
  depth: natural := 16;
  addr_width: natural := 4); -- log depth to the base 2
  port (addr: in std_logic_vector(addr_width-1 downto 0);
        datain: in std_logic_vector(width-1 downto 0);
        dataout: out std_logic_vector(width-1 downto 0);
        rw, clk: in std_logic);
end ram_vhdl;

architecture behv of ram_vhdl is

subtype wtype is std_logic_vector(width-1 downto 0);
type mem_type is array(depth-1 downto 0) of wtype;
signal memory : mem_type;

begin
  process
  begin
    wait until clk = '1' and clk'event;
    if (rw = '0') then -- write
      memory(conv_integer(addr)) <= datain;
    end if;
  end process;
  process(rw, addr)
  begin
    if (rw = '1') then -- read
      dataout <= memory(conv_integer(addr));
  else
      dataout <= wtype'(others => 'Z');
```

```
    end if;
  end process;
end behv;
```

**Verilog Code**

```
// One Port Synchronous RAM with Read/Write Enable

module ram (addr, datain, dataout, rw, clk);

parameter width = 8,
              depth = 16,
              addr_width = 4; // log depth to the base 2

input [addr_width-1:0] addr;
input [width-1:0] datain;
output [width-1:0] dataout;
input clk;
input rw;

reg [width-1:0] dataout;
reg [width-1:0] memory [depth-1:0];

always @(posedge clk)
 begin
  if (!rw) // write
     memory[addr] = datain;
 end
always @(rw)
 begin
  if (rw) // read
     dataout = memory[addr];
  else
     dataout = 8'bz;  // Verilog does not allow (width-1)'bz
 end
endmodule
```

### Example 2.13 HDL Description of a ROM

```
// Modeling of ROMs

module rom(data, address);

input[9:0] address;
output[7:0] data;

/* synopsys translate_off */

reg[7:0] memory[0:1023];

initial
begin
  $readmemh("/home/rom.data", memory);
end

assign data = memory[address];
/* synopsys translate_on */
endmodule
```

### Example 2.14 HDL Description of an Incrementor

**VHDL Code**

```
library IEEE;
use IEEE.std_logic_1164.all;
use IEEE.std_logic_unsigned.all;

entity inc_vhdl is
  generic(width : integer := 32);
  port ( datain: in std_logic_vector(width-1 downto 0);
         control: in std_logic;
         dataout: out std_logic_vector(width-1 downto 0);
         flag: out std_logic);
end inc_vhdl;

architecture behv of inc_vhdl is
```

```
signal dataout_int: std_logic_vector(width-1 downto 0);

begin

  process(datain, control)
  begin
    if (control = '1') then -- increment
      dataout_int <= datain + '1';
    else -- feedthrough
      dataout_int <= datain;
    end if;
  end process;

  dataout <= dataout_int;

  flag <= '1' when (control = '1' and datain = To_stdlogicvector(X"FFFF")) else '0';

end behv;
```

**Verilog Code**

```
`timescale 1 ns / 1 fs
module incrementer (datain, control, dataout, flag);

parameter width = 32;  // parametrizable width

input [width-1:0] datain;
output [width-1:0] dataout;
input control;
output flag;

parameter INCR = 1;
parameter FTHRU = 0;  // Feedthrough

reg [width-1:0] dataout;

always @(datain or control) begin
  if (control == INCR)
      dataout = datain + 1'b1;
  else
```

```
      dataout = datain;
end // always

assign flag = (~(|dataout) && control) ? 1 : 0;

endmodule
```

Example 2.14 shows the VHDL and Verilog code for a simple combinational logic block. It is a parametrizable incrementor which can be programmed in either a feedthrough mode or the increment mode. Also there is a flag logic which indicates when the design counts to 0 in the incrementor mode.

## Example 2.15 HDL Description of a Barrel Shifter

### VHDL Code

```
library IEEE;
use IEEE.std_logic_1164.all;
use IEEE.std_logic_arith.all;

entity bs_vhdl is
  port (datain: in std_logic_vector(31 downto 0);
  direction: in std_logic;
  count: in std_logic_vector(4 downto 0);
  dataout: out std_logic_vector(31 downto 0));
end bs_vhdl;

architecture behv of bs_vhdl is

function barrel_shift(din: in std_logic_vector(31 downto 0);
  dir: in std_logic;
  cnt: in std_logic_vector(4 downto 0)) return std_logic_vector is
begin

  if (dir = '1') then
     return std_logic_vector((SHR(unsigned(din), unsigned(cnt))));
  else
    return std_logic_vector((SHL(unsigned(din), unsigned(cnt))));
  end if;
end barrel_shift;
```

```
begin

  dataout <= barrel_shift(datain, direction, count);

end behv;
```

**Verilog Code**

```
`timescale 1 ns / 1 fs

module bs_ver (datain, count, direction, dataout);

input [31:0] datain;
input [4:0] count;
input direction;  // RIGHT or LEFT
output [31:0] dataout;

parameter RIGHT = 1;
parameter LEFT = 0;

function [31:0] barrel_shift;
input [31:0] din;
input dir;
input [4:0] cnt;
begin
  barrel_shift = (dir == RIGHT) ? (din >> count) : (din << count);
end
endfunction

  assign dataout = barrel_shift(datain, direction, count);
endmodule
```

Example 2.15 describes a barrel shifter which is capable of shift left or shift right based upon an input control signal. The width of the data bus is 32 bits and the shift count bus is 5 bits wide. Zeros are shifted in when shifting either right or left.

## Example 2.16 HDL Description of a Multi-Bit Register

### VHDL Code

```
library IEEE;
use IEEE.std_logic_1164.all;

entity reg_p is
  generic (width : natural := 8);
  port (r : in std_logic_vector(width-1 downto 0);
        clk, ena, rst: in std_logic;
        data: out std_logic_vector(width-1 downto 0));
end reg_p;

architecture behv of reg_p is

signal gclk : std_logic;

begin

  gclk <= clk and ena;

  process(rst, gclk)
  begin

   if (rst = '0') then
       data <= (others => '0');
   elsif gclk'event and gclk = '1' then
       data <= r;
   end if;
  end process;

end behv;
```

### Verilog Code

```
`timescale 1 ns / 1 fs
module register (r, clk, data, ena, rst);
```

```
parameter width = 8;

input [width-1:0] r;
output [width-1:0] data;
input clk, ena, rst;

reg [width-1:0] data;
and I1 (gclk, clk, ena);

always @(posedge gclk or negedge rst)
begin
  if (rst == 1'b0) data = 0;
  else data = r;
end
endmodule
```

Example 2.16 describes a parametrizable multi bit register with asynchronous reset and clock enable signal.

## 2.5 Classic Scenarios

This section discusses classic coding scenarios for DC. Unlike the classic scenarios discussed in chapter 1, these are useful to all designers writing HDL code for synthesis using DC.

**Scenario 1**

You wish to infer logic which models a parity detector. In other words, you have an incoming databus, whose parity when odd, returns a 1, and a 0 when parity is even. In order to reuse the code irrespective of the bus length, you wish to write a function and use the function to infer the logic.

**Solution**

The following example shows one way of coding the above, to realize the parity detector.

**VHDL Code**

```
-- This is an example of a synthesizable parametrizable Parity Detector
library IEEE;
use IEEE.std_logic_1164.all;
entity par is
```

```
  generic(width: natural := 8);
  port(data: in std_logic_vector((width-1) downto 0);
        parity:out std_logic);
end par;
architecture behv of par is
function cal_parity(datain: in std_logic_vector) return std_logic is
-- temporary variable to count the number of 1's
variable count: integer range datain'range;
begin
  count := 0;
  for i in datain'range loop
    if (datain(i) = '1') then
      count := count + 1;
    end if;
  end loop;
  if (count mod 2) = 1 then
    return '1';
  else
    return '0';
  end if;
  end cal_parity;
  begin
     parity <= cal_parity(data);
end behv;
```

### Verilog Code

```
/* This is an example of a synthesizable parametrizable Parity Detector
Note the modulus operator requires constant operands in verilog. Hence cannot be
used */
module parity (data, parity);
parameter width = 8;
input [width-1:0] data;
output parity;
function cal_parity;
input [width-1:0] datain;
integer i;
reg int_parity;
begin
  int_parity = 1'b0;
```

```
  for (i = 0; i <= width-1; i = i + 1) begin
     if (datain[i] == 1'b1) begin
       if (int_parity == 1'b0)
          int_parity = 1'b1;
       else
          int_parity = 1'b0;
       end
     end
     cal_parity = int_parity;
  end
endfunction
   assign parity = cal_parity(data);
endmodule
```

**Scenario 2**

How does one code a mod N counter with terminal count.

**Solution**

Shown below is an example in VHDL and Verilog which, when synthesized gives a mod N counter with terminal count.

**VHDL Code**

```
library IEEE;
use IEEE.std_logic_1164.all;
entity counter is
  generic (N: natural := 3); -- N is a top level generic
  port (clear: in std_logic;
      clock: in std_logic;
     count: out integer range 0 to (N - 1));
end counter;
architecture behv of counter is
signal count_int : integer range 0 to (N-1) ;
begin
  count <= count_int ;
  process
  begin
    wait until clock = '1' and clock'event;
    if (clear = '1' or count_int >= (N - 1)) then
      count <= 0;
```

```
      else
          count_int <= count_int + 1;
      end if;
  end process;
end behv;
```

**Verilog Code**

```
module modNcounter (clear, clock, count);
parameter N = 3;
parameter width = 2; /* minimum number of bits required */
input clear;
input clock;
output [width-1 : 0] count;
reg [width-1:0] count_int;
assign count = count_int;
always @(posedge clock) begin
  if ((clear == 1'b1) | (count_int >= (N - 1)))
      count_int = 0;
  else
      count_int = count_int + 1;
end
endmodule
```

**Scenario 3**

An eight bit counter consisting of two groups of four bits each, has two load signals Load1 and Load2 as shown in the Figure 2-7. There is also an eight bit bus. When Load1 is active, the lower four bits of the bus are loaded into the lower four bits of the counter. Similarly, when Load2 is active, the higher four bits are loaded into the counter. If neither is active, the counter performs the count operation. The figure does not show the additional combinational logic that is required. How does one code for this?

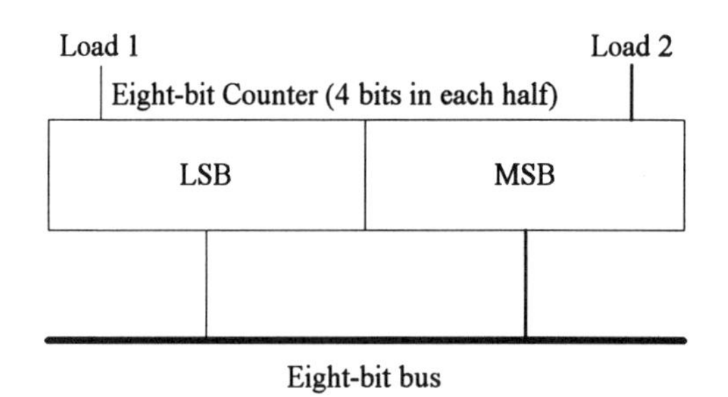


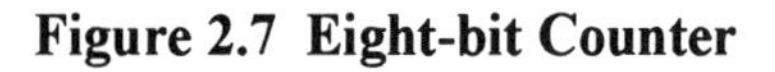

**Figure 2.7 Eight-bit Counter**

**Solution**

Shown below is one way of coding for the load counter.

**VHDL Code**

```
library ieee;
use ieee.std_logic_1164.all;
use ieee.std_logic_unsigned.all;
entity ld_counter is
  port (load: in std_logic_vector(1 downto 0);
         databus : in std_logic_vector(7 downto 0);
          clk: in std_logic;
          count : out std_logic_vector(7 downto 0));
end ld_counter;
architecture behv of ld_counter is
signal int_count: std_logic_vector(7 downto 0);
begin
   count <= int_count;
   process
   begin
    wait until clk = '1' and clk'event;
    if (load = "10") then
         int_count <= databus(7 downto 4) & int_count(3 downto 0);
    elsif (load = "01") then
         int_count <= int_count(7 downto 4) & databus(3 downto 0);
    else
         int_count <= int_count + 1;
```

```
    end if;
  end process;
```

**Verilog Code**

```
module ld_cntr (load, databus, clk, count);
input [1:0] load;
input [7:0] databus;
input clk;
output [7:0] count;
reg [7:0] int_count;
always @(posedge clk) begin
  if (load == 2'b10)
      int_count = {databus[7:4],int_count[3:0]};
  else if (load == 2'b01)
      int_count = {int_count[7:4], databus[3:0]};
  else
      int_count = int_count + 1;
end
assign count = int_count;
endmodule
```

**Scenario 4**

You are trying to infer shift registers using Verilog HDL. Instead of getting registers connected one after the other, you find two registers each connected to the same inputs!

**Solution**

When infering shift registers using Verilog HDL, the use of blocking or non-blocking RTL assignments will determine the kind of results that are generated. When blocking assignments are used, the order of the assignments specified is important since each assignment is executed before proceeding to the next statement. For example, if it is necessary to create a shift register with reg A connected to reg B, the following code must be used if blocking assignments are used:

```
always@clk
begin
  B = A;
  A = C;   /* C is an input */
end
```

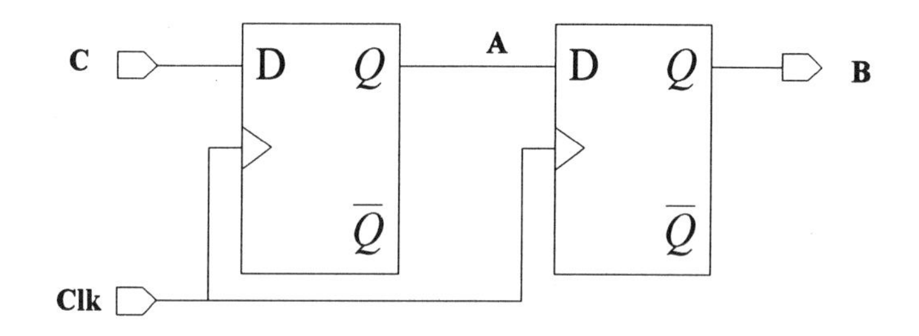


**Figure 2.8 Desired Two Flop Structure**

If the following code is used instead, both reg A and B will be tied to the same input C:

```
always @ posedge clk
begin
   A = C;
   B = A;
end;
```

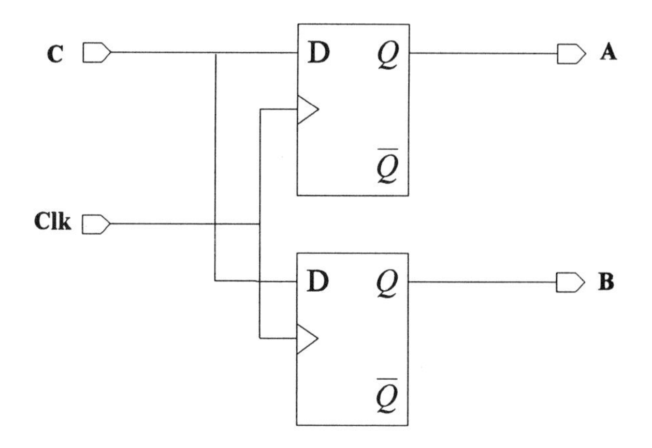


**Figure 2.9 Faulty two flop structure**

When non-blocking RTL assignments are used, the order is not important as long as the intended functionality is conveyed. Thus both of the following examples will produce the same results:

```
always@ posedge clk
begin
      A <= C;    B <= A;
end
always@ posedge clk
begin
  B <= A ;
  A <= C ;
end
```

**Scenario 5**

How does one code to infer an updown load counter on synthesis?

**Solution**

Shown below is an example in both Verilog and VHDL to synthesize an updown load counter on synthesis.

**Verilog Code**

```
module updnctr(rst,load,pre,updn,clk,din,cnt);
 input rst,load,pre,updn,clk;
 input [7:0] din;
 output [7:0] cnt;
 reg [7:0] cnt;
always @(posedge clk)
begin
      if (rst==1)
         cnt = 8'h00; else if (pre==1) cnt = 8'h11; else if (load==1)
         cnt = din;
      else if (updn==1) cnt = cnt + 1;
      else if (updn==0) cnt = cnt - 1;
end
endmodule
```

**VHDL Code**

```
library IEEE;
use IEEE.std_logic_1164.all;
use IEEE.std_logic_unsigned.all;
entity up_dn_ctr is
```

```
port (rst, load, pre, updn, clk: in std_logic;
  din: in std_logic_vector(7 downto 0);
  cnt: out std_logic_vector(7downto 0));
end up_dn_ctr;
architecture behv of up_dn_ctr is
signal int_cnt: std_logic_vector(7 downto 0);
begin
process
begin
  wait until clk = '1' and clk'event;
   if    (rst = '1')    then int_cnt <= "00000000";
   elsif (pre = '1') then int_cnt <= "11111111";
   elsif (load = '1') then int_cnt <= din;
   elsif (updn = '1') then int_cnt <= int_cnt + 1;
   else int_cnt <= int_cnt - 1; -- (updn = '0') then
   end if;
end process;
cnt <= int_cnt;
end behv;
```

**Scenario 6**

You wish to read a variable length string from a file in VHDL. How does one code this in VHDL since the READ procedure in the TEXTIO package requires a constrained length string parameter.

**Solution**

Here is one way of doing this purely for simulation purposes and not for synthesis. For a variable L of type Line, attribute L'Length gives the number of characters on that line remaining to be read.

Here is the relevant VHDL code section

```
   -- Loop continuously till end of file is encountered
   while not (endfile(Reader)) loop
-- Reading a line from the file pointed by "READER"
      readline(Reader, L);
      wait for 5 ns;
      size := L'Length; -- Computing the size of the string to be read
      temp1 := (others => NUL); -- Initializing the string variable
      read(L, temp1(size downto 1));
   end loop;
```

## Recommended Further Readings

1. *VHDL Compiler Reference Manual*
2. *HDL Compiler for Verilog Reference Manual*
3. *HDL Coding Styles: Sequential Devices Application Note*
4. Synopsys Newsletter *Impact* Support Center Q&A

CHAPTER 3

# *Pre and Post-Synthesis Simulation*

Simulation is the process of verifying the functionality and timing of a design against its original specifications. In the ASIC design flow, designers perform functional simulation prior to synthesis. After synthesis, gate level simulation is performed on the netlist generated by synthesis. This chapter has been included to provide a better understanding of the synthesis-based ASIC design flow. Since the focus of this book is primarily synthesis, this chapter does not delve into details of either simulation or the simulation tool used. The simulator used is the Synopsys *VHDL System Simulator (VSS)*.

In this chapter, we discuss an example VHDL code, provide a testbench for the same example and the output files of the simulation performed using VSS. Two approaches to simulation are discussed. The first approach is suited to an interactive simulation with waveforms, while the second involves reading the stimulus from a test file and writing the output to another file.

## 3.1 RTL Simulation

VHDL RTL simulation is used to verify that the design coded in VHDL captures the functionality required by the design specifications. After the design has been coded in VHDL, a testbench which applies stimulus to the design must be created. The design to be simulated is referred to as the Design Under Test (DUT). The outputs generated by the DUT can then be compared in the testbench (a smart testbench) or verified interactively during debugging.

### 3.1.1 TAP Controller

The Joint Test Action Group (JTAG) Boundary Scan Standard was initiated to ensure that all ASICs have a common denominator of DFT circuitry which will make the test development and board level testing a standard. This standard allows for efficient testing of board interconnect and facilitates isolation and testing of chips. The TAP controller is part of the control logic that resides on the chip to interface the JTAG bus to the DFT hardware residing on the chip. Detailed description of the TAP controller architecture and functionality are beyond the scope of this book.

The TAP controller is a synchronous finite-state machine whose state diagram is shown in Figure 3.1. It has a single input TMS and its outputs are signals corresponding to different states of the state-machine. Example 3.1 shows the VHDL description of the TAP controller. The TAP controller described conforms to the 1149.1 standard.

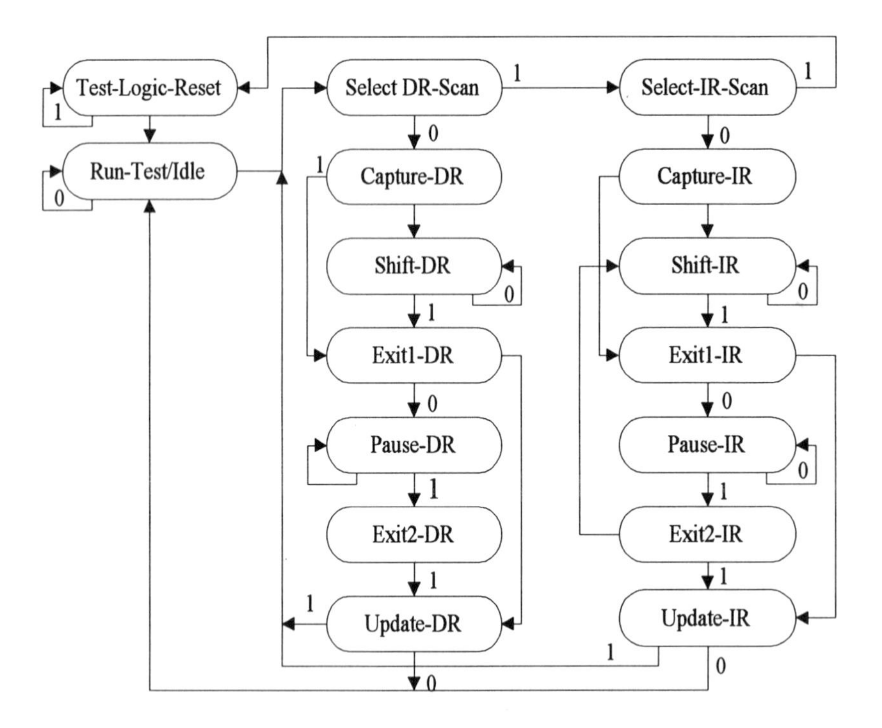


**Figure 3.1 State Diagram for TAP Controller**

## Example 3.1 Code for TAP Controller

### VHDL Code

```
package pack is
-- enumerated type showing all states of the state machine
type tap_states is (TLR, RTI, Sel_DR_Scan, Cap_DR, Shft_DR, Exit1_DR,
Pause_DR, Exit2_DR, Update_DR, SEl_IR_Scan, Cap_IR, Shft_IR, Exit1_IR,
Pause_IR, Exit2_IR, Update_IR);
end pack;
library IEEE;
use IEEE.std_logic_1164.all;
use work.pack.all;
entity tap_controller is
  port (TMS: in std_logic;
      TCK: in std_logic;
       tstate: out tap_states;
       TRST: in std_logic);
end tap_controller;
architecture behv of tap_controller is
signal pstate, nstate: tap_states;
begin
combo: process(pstate, tms)
  begin
  case (pstate) is
when TLR =>
 if (tms = '1') then
               nstate <= pstate;
             else
               nstate <= RTI;
             end if;
when RTI =>
 if (tms = '1') then
               nstate <= Sel_DR_Scan;
             else
               nstate <= pstate;
             end if;
when Sel_DR_Scan =>
 if (tms = '1') then
               nstate <= Sel_IR_Scan;
```

```
                else
                  nstate <= Cap_DR;
                end if;
when Cap_DR =>
 if (tms = '1') then
                  nstate <= Exit1_DR;
                else
                  nstate <= Shft_DR;
                end if;
when Shft_DR =>
 if (tms = '1') then
                  nstate <= Exit1_DR;
               else
                  nstate <= pstate;
               end if;
when Exit1_DR =>
 if (tms = '1') then
                  nstate <= Update_DR;
               else
                  nstate <= Pause_DR;
               end if;
when Pause_DR =>
 if (tms = '1') then
                  nstate <= Exit2_DR;
                else
                  nstate <= pstate;
                end if;
when  Exit2_DR =>
 if (tms = '1') then
                  nstate <= Update_DR;
                else
                  nstate <= Shft_DR;
                end if;
when Update_DR =>
 if (tms = '1') then
                  nstate <= Sel_DR_Scan;
                else
                  nstate <= RTI;
                end if;
```

```
when Sel_IR_Scan =>
 if (tms = '1') then
               nstate <= TLR;
             else
               nstate <= Cap_IR;
             end if;
when Cap_IR =>
 if (tms = '1') then
               nstate <= Exit1_IR;
             else
               nstate <= Shft_IR;
             end if;
when Shft_IR =>
 if (tms = '1') then
               nstate <= Exit1_IR;
             else
               nstate <= pstate;
             end if;
when Exit1_IR =>
 if (tms = '1') then
               nstate <= Update_IR;
             else
               nstate <= Pause_IR;
             end if;
when Pause_IR =>
 if (tms = '1') then
               nstate <= Exit2_IR;
             else
               nstate <= pstate;
             end if;
when  Exit2_IR =>
 if (tms = '1') then
               nstate <= Update_IR;
             else
               nstate <= Shft_IR;
             end if;
when Update_IR =>
 if (tms = '1') then
               nstate <= Sel_DR_Scan;
```

```
                else
                  nstate <= RTI;
                end if;
end case;
end process combo;
seq: process(TRST, TCK)
begin
if (TRST = '0') then
  pstate <= TLR;
elsif (TCK = '1') and (TCK'event) then
  pstate <= nstate;
end if;
end process seq;
  tstate <= pstate;
end behv;
```

### 3.1.2 Testbench for TAP Controller

In this section, a testbench for the TAP controller discussed in section 3.1.1 is shown. Stimulus is directly applied in the VHDL code. The outputs can be viewed in a waveform viewer provided with any VHDL simulator.

### Example 3.2 Testbench for TAP Controller

```
library IEEE;
use IEEE.std_logic_1164.all;
use work.pack.all;
entity test is
end;
architecture test1 of test is
  signal clock, control, reset: std_logic := '0';
  signal stateout: tap_states := TLR;
  component tap_controller
  port (TMS: in std_logic;
        TCK: in std_logic;
        tstate: out tap_states;
        TRST: in std_logic);
  end component;
 for U1: tap_controller use entity work.tap_controller(behv);
```

```
begin
 U1: tap_controller port map (control,clock,stateout,reset);
 clock_tic: process begin
  loop
    clock <= '1';
    wait for 5 ns;
    clock <= '0';
    wait for 5 ns;
  end loop; end process;

 state_changes: process
  begin
    control <= '1' after 5 ns, '0' after 55 ns,
            '1' after 75 ns, '0' after 85 ns,
            '1' after 125 ns, '0' after 135 ns,
            '1' after 155 ns, '0' after 185 ns,
            '1' after 195 ns, '0' after 225 ns,
            '1' after 235 ns, '0' after 245 ns,
            '1' after 255 ns, '0' after 265 ns,
            '1' after 275 ns, '0' after 295 ns,
            '1' after 305 ns, '0' after 325 ns,
            '1' after 355 ns, '0' after 365 ns,
            '1' after 385 ns, '0' after 395 ns,
            '1' after 405 ns, '0' after 425 ns,
            '1' after 435 ns, '0' after 455 ns,
            '1' after 465 ns, '0' after 475 ns,
            '1' after 485 ns;
    wait;
  end process;
  reset <= '1';
end TEST1;
```

**Simulation Output**

Synopsys 1076 VHDL Simulator

```
0 NS
    M:   EVENT /TEST/CLOCK (value = '1')
   M2:   EVENT /TEST/RESET (value = '1')
   M3:   EVENT /TEST/STATEOUT (value = RTI)
5 NS
```

```
    M1:   EVENT /TEST/CONTROL (value = '1')
     M:   EVENT /TEST/CLOCK (value = '0')
10 NS
     M:   EVENT /TEST/CLOCK (value = '1')
    M3:   EVENT /TEST/STATEOUT (value = SEL_DR_SCAN)
```

## 3.2 File Text IO in VHDL Using the TEXTIO Package

In section 3.1.2, we discussed a testbench where the stimulus was specified in the VHDL code. In this section, we discuss an alternative approach to VHDL simulation. This requires the use of the TEXTIO from a TEXTIO file. In this approach, stimulus is read from a stimulus file and the outputs of the DUT are written to an output file.

To perform TEXTIO in VHDL, one must use the TEXTIO package supplied in the STD library. The TEXTIO package contains declarations of types and subprograms that support formatted ASCII I/O operations. The Std_logic_Textio package overloads these I/O subprograms for the std_logic base types.

### Example 3.3 Testing TAP Controller Using TEXTIO

```
package pack is
type tap_states is (TLR, RTI, Sel_DR_Scan, Cap_DR, Shft_DR, Exit1_DR,
Pause_DR, Exit2_DR, Update_DR, SEl_IR_Scan, Cap_IR, Shft_IR, Exit1_IR,
Pause_IR, Exit2_IR, Update_IR);

-- This function takes in the enumerated type tap_states and
-- returns type STRING
-- The function converts the enumerated type tap_states defined above to
--string so that one can then use the “write” procedure from the textio
--package.
function to_string(A: in tap_states) return string;
end pack;
package body pack is
function to_string(A: in tap_states) return string is
begin
  case A is
  when TLR => return “TLR”;
  when RTI => return “RTI”;
  when Sel_DR_Scan => return “Sel_DR_Scan” ;
  when Cap_DR => return “Cap_DR”;
  when Shft_DR => return “Shft_DR”;
```

```
  when Exit1_DR => return "Exit1_DR";
  when Pause_DR => return "Pause_DR";
  when Exit2_DR => return "Exit2_DR";
  when Update_DR => return "Update_DR";
  when SEI_IR_Scan => return "SEI_IR_Scan";
  when Cap_IR => return "Cap_IR";
  when Shft_IR => return "Shft_IR" ;
  when Exit1_IR => return "Exit1_IR";
  when Pause_IR => return "Pause_IR";
  when Exit2_IR => return "Exit2_IR";
  when Update_IR => return "Update_IR";
  end case;
end to_string;
end pack;
-- This VHDL testbench shows a template for testing the TAP Controller
-- using File TEXTIO
-- Make declarations and subprograms in the package TEXTIO
--visible
use std.textio.all;
library IEEE;
use IEEE.std_logic_1164.all;
-- std_logic_textio package overloads the procedures in
-- the package
--TEXTIO for std_logic type
use IEEE.std_logic_textio.all;
-- Make the type tap_states visible
use work.pack.all;
entity testbench is
end testbench;

architecture file_io of testbench is
-- Defining a file object "READER" of mode IN and type TEXT
file READER: TEXT is in "DATAIN";
file WRITER: TEXT is out "DATAOUT";
component tap_controller
port (TMS: in std_logic;
     TCK: in std_logic;
     tstate: out tap_states;
     TRST: in std_logic);
```

```
end component;
-- Declaring temporary signals for interfacing with ports
signal TMS, TCK, TRST: std_logic;
signal tstate: tap_states;
begin
  UUT: tap_controller port map (TMS, TCK, tstate, TRST);
-- Clock generation
  clock_tic: process
  begin
      TCK <= '1';
      wait for 5 ns;
      TCK <= '0';
       wait for 5 ns;
if (now > 400 ns) then
        wait;
end if;
 end process;
-- This process reads the stimulus from a file and applies
-- it to the primary input ports
  P1: process
-- Defining the local variable L of type line
  variable L: line;
-- The variable duration is used to determine the time
-- when data needs to be applied to the port inputs
  variable duration: time;
  variable test_mde_select, test_reset: std_logic;
  begin
-- Add this line if there are comments at the top of
-- the file "DATAIN"
readline(Reader, L);
-- Loop continuously till end of file is encountered
while not (endfile(Reader)) loop
-- Reading a line from the file DATAIN
readline(Reader, L);
  read(L, duration);
  wait for (duration - now);
  read(L, test_mde_select);
  read(L, test_reset);
  TMS <= test_mde_select;
```

```
  TRST <= test_reset;
      end loop;
      assert (false) report "Simulation Complete. Thanks"
      severity note;
       wait;
  end process P1;
  P2: process
  variable L: line;
  begin
    wait until TCK = '1' and TCK'event;
    write(L, now, right, 2);
    write(L, to_string(tstate), right, 20);
    writeline(WRITER, L);
      if (now > 400 ns) then
       wait;
      end if;
end process P2;
end file_io;
configuration CFG_top of testbench is
for file_io
 for all: tap_controller use entity work.tap_controller(behv);
 end for;
end for;
end CFG_top;
```

**Input Stimulus File**

```
TIME  TMS TRST
5 ns    1   0
15 ns   0   1
25 ns   1   1
35 ns   0   1
45 ns   0   1
55 ns   1   1
65 ns   0   1
75 ns   1   0
85 ns   0   1
95 ns   1   1
105 ns  0   1
115 ns  0   1
```

```
125 ns  1  1
135 ns  0  1
145 ns  1  0
```

**Simulation Output Results**

```
0 NS        TLR
10 NS       RTI
20 NS       RTI
30 NS       RTI
40 NS       Sel_DR_Scan
50 NS       Cap_DR
60 NS       Shft_DR
70 NS       Exit1_DR
80 NS       TLR
90 NS       TLR
100 NS      RTI
110 NS      Sel_DR_Scan
120 NS      Cap_DR
130 NS      Shft_DR
140 NS      Exit1_DR
150 NS      TLR
160 NS      TLR
170 NS      TLR
180 NS      TLR
190 NS      TLR
200 NS      TLR
210 NS      TLR
220 NS      TLR
230 NS      TLR
240 NS      TLR
250 NS      TLR
260 NS      TLR
270 NS      TLR
280 NS      TLR
290 NS      TLR
```

## 3.3 VHDL Gate Level Simulation

To perform gate level simulation of a VHDL netlist one requires the VHDL simulation libraries from the ASIC vendor. The Synopsys liban utility can generate the VHDL library models from the synthesis technology library. For the more complex cells, simulation models will have to be manually created. The VHDL models generated are encrypted so that the vendor proprietary information is protected.

## 3.4 Verilog Gate Level Simulation

This section provides an example of a testbench in verilog. The instantiation of the UUT (Unit Under Test), generation of clock stimuli and ending the simulation run at a predefined time are illustrated.

### Example 3.4 Verilog Testbench

```
`resetall
`timescale 1 ns / 1 fs

module testfixture;
parameter top_width = 10;
wire [top_width-1:0] data;
reg [top_width-1:0] r;
reg clk, ena, rst;

`define stimuli #10 r =
/* Instantiation of Unit Under Test */

register #(top_width) uut (.r(r), .clk(clk), .data(data), .ena(ena), .rst(rst));
initial
begin
  clk = 0;
  rst = 1; // reset inactive at start
  ena = 1;
  r = 10;
end

// clock stimuli
always
begin
```

```
#5 clk = ~clk;
if ($time == 100) $finish;
end
initial
begin
  #3 ena = 1'b0; rst = 1'b0;
  #12 rst = 1'b1;
  #8 ena = 1'b1;
  `stimuli 6;
  `stimuli 8;
  `stimuli 10;
  `stimuli 12;
end
always @(posedge clk)
$strobe("at time %0d r = %d data = %d ena = %b rst = %b", $time, r, data, ena, rst);
endmodule
```

## 3.5 Classic Scenarios

In this section, we discuss a few basic simulation related classic scenarios. The simulator used is Synopsys VSS.

**Scenario 1**

You are running the simulation for several time periods but you still do not see any waveforms.

**Solution**

Default timebase is femtoseconds in VSS. Check to see if you have specified your timebase as ns or any larger timebase instead of the default.

**Scenario 2**

When analyzing the configuration of a VHDL design with generate statements, VSS issues the following warning.

```
for all: COMP use entity work.comp_ent(behav);
Warning: vhdlan,782 generate.vhd(45):
Component is never instanced in the corresponding set of statements.
"generate.vhd": errors: 0; warnings: 1.
```

**Solution**

This warning is most likely due to the configuration statement. Here is an example of a VHDL code using nested generate statements.

```
library IEEE;
use IEEE.std_logic_1164.all;
entity generate_eg  is
  generic(N : natural := 6);
   port (A : in std_logic_vector((N-1) downto 0);
        B : out std_logic_vector((N-1) downto 0));
end generate_eg;
architecture usage of generate_eg is
   component COMP
        port (X : in std_logic;
             Y : out std_logic);
   end component;

begin
    GEN : for I in (N-1) downto 0 generate
        GEN1: if (I = 3) generate
              U: COMP port map( X => A(I), Y => B(I));
        end generate;
   end generate;
end usage;
```

The configuration statement required to configure this design is as follows:

```
configuration CFG_generate_eg of generate_eg is
   for usage
     for GEN((N-1) downto 0)
        for GEN1
          for all: COMP use entity work.comp_ent(behav);
          end for;
        end for;
     end for;
   end for;
end CFG_generate_eg;
```

In this example, there is a top level block configuration (starting for usage) within the configuration declaration for the top level architecture usage. Within this block configuration is a configuration item (starting for GEN) which opens up the scope of the outer generate statement GEN in the architecture. This makes names defined within it, visible from within the for usage statement. Since our design has nested generate statements, we need another block configuration nested within it (starting for GEN1). This opens up the scope of the component instantiation within it, and hence we have a final component configuration which binds the component COMP to a design unit entity comp_ent, and design unit architecture behav within the library work.

**Scenario 3**

You are performing gate level simulation. You have analyzed the VHDL gate level simulation models into a library. On invoking the simulator, you get a message that components are unbound.

**Solution**

To avoid the unbound components message, follow the steps given below:

1. Define a VHDL design library gate_lib, where all the simulation models for the gates are available.
2. Ensure that there is a library gate_lib clause declared before the top level entity to be simulated. If you have packages, make sure the library clause is immediately before the entity.
3. Analyze the simulation models and the components package to the library gate_lib.
4. Make sure there is a top level configuration such as the one shown below:
   ```
   configuration CFG_top of count_seq_vhdl is
   for SYN_behavior
   end for;
   end CFG_top;
   ```
5. Analyze design netlist to library work.
6. Invoke simulator on the top level configuration.

**Scenario 4**

You have performed behavioral simulation of your code and all your functional vectors pass through fine. After synthesizing the design to gates using DC, you begin gate level simulation. During this simulation, you activate the synchronous reset line to reset all your flops in your design to a known value. You find that the reset is not the controlling signal in the design and does not propagate a known value to the data pins of the flops.

**Solution**

A possible scenario for this problem could be the case where the output of a flip-flop could be driving the select line of a mux, and the mux output drives the data input of the flip-flop as shown in Figure 3-2.

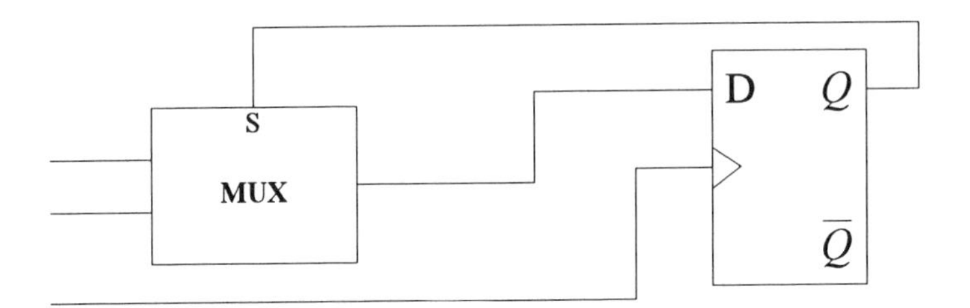


**Figure 3.2 Possible Scenario for "unknown" at Flop Data Input**

At the start of gate-level simulation all the sequential elements in the netlist are uninitialized. In VHDL they are all "U"s and in Verilog "X"s. If the design has a synchronous reset and goes through several levels of logic ie.it is not the controlling signal, then the uninitialized flops might prevent the propagation of the reset value during gate level simulation. One such scenario is shown in the figure. Since the flop is uninitialized at the start, the synchronous reset value will not propagate through the mux. One option is to manually initialize all the sequential elements in the design to an arbitrary known value at the start of gate level simulation.

The other alternative is to identify the synchronous reset signal to the logic synthesis tool such that it synthesizes the reset logic as close to the data pin of the sequential elements as possible so as to make it the controlling signal. In DC this is possible using the sync_set_reset attribute. There might be some timing overhead involved to achieve this. Shown below is an example showing the use of the sync_set_reset attribute.

**VHDL Code**

```
attribute sync_set_reset of RESET : signal is "true";
process(RESET, CLK)
begin
  if (CLK'event and CLK = '1') then
    if RESET = '1' then
      Q <= '0';
    else
      Q <= DATA_A;
```

```
    end if;
  end if;
end process;
```

**Verilog Code**

```
// synopsys sync_set_reset "reset"

reg y;
always @ (posedge clk)
begin : synchronous_reset
   if (reset)
     y = 1'b0; --synchronous reset
   else
     y = d1;
end
```

**Scenario 5**

Is it possible to perform textio from the computer keyboard. In other words is possible to write VHDL and simulate such that the user is prompted for the stimulus?

**Solution**

This is possible in VSS, provided the following variable is set to true in the .synopsys_vss.setup file.

```
PROMPT_STD_INPUT = TRUE
```

Shown below is an example in VHDL to do the same.

**VHDL Code**

```
use std.textio.all;
library IEEE;
use IEEE.std_logic_textio.all;
entity keyboard_textio is
end keyboard_textio;
architecture behv of keyboard_textio is
-- This is already defined in the TEXTIO package
-- file INPUT: TEXT is in "STD_INPUT";
-- file OUTPUT: TEXT is out "STD_OUTPUT";
begin
process
```

```
variable number : integer;
variable line_ptr: line;
variable flag: boolean := false;
begin
      while (not flag) loop
        write(line_ptr, string'("Enter the data"));
        writeline(OUTPUT, line_ptr);
        readline(INPUT, line_ptr);
        read(line_ptr, number, flag);
      end loop;
      wait for 10 ns;
end process;
end behv;
```

**Scenario 6**

You are generating FTGS VHDL simulation models from your Synopsys synthesis library. Is there a way to turn off setup and hold checks on the scan path for multiplexed flipflop scan cells in the library when the scan_enable is inactive, that is, when the scan path is not selected.

**Solution**

The when and sdf_cond attributes in Library Compiler can be used to achieve this. Here is an example showing the relevant section of the library.

```
 pin(TI) {
   direction : input;
   capacitance : 1;
   timing() {
     when : "TE";
     sdf_cond : "b == 1'TE1" ;
     timing_type : setup_rising;
     intrinsic_rise : 1.3;
     intrinsic_fall : 1.3;
     related_pin : "CP";
   }
   timing() {
     when : "TE";
     sdf_cond : "b == 1'TE1" ;
     timing_type : hold_rising;
```

```
    intrinsic_rise : 0.3;
    intrinsic_fall : 0.3;
    related_pin : "CP";
  }
}
```

## Recommended Further Readings


1. *VHDL: Hardware Description and Design*, Roger Lipsett, Carl Schaefer, Cary Ussery.
2. *The Verilog Hardware Description Language, Second Edition,* Donald E. Thomas, Philip R. Moorby.

CHAPTER 4

# *Constraining and Optimizing Designs - I*

This chapter is the first of two chapters that outlines the methodology for logic synthesis using the Synopsys *Design Compiler.* After a design has been described in HDL and functionally simulated, the next step involves logic synthesis using DC. Herein lies the core of the synthesis process. How can one get the best results from the synthesis tool? What is the methodology to be followed in optimizing a design? Is synthesis a push-button solution? This chapter begins with a description of the constraints specified on designs and the timing reports generated by DC. A brief description of commonly used DC commands and options has been provided to help the reader get familiar with the process of optimizing designs. Then, strategies for optimization and general guidelines for synthesis are discussed. Finally, a number of "classic scenarios" have been presented based on actual user experiences. At each stage, the relevant dc_shell commands have been provided.

## 4.1 Synthesis Background

The DC attempts to meet two basic constraints, namely:

1. Design rule constraints
2. Optimization constraints

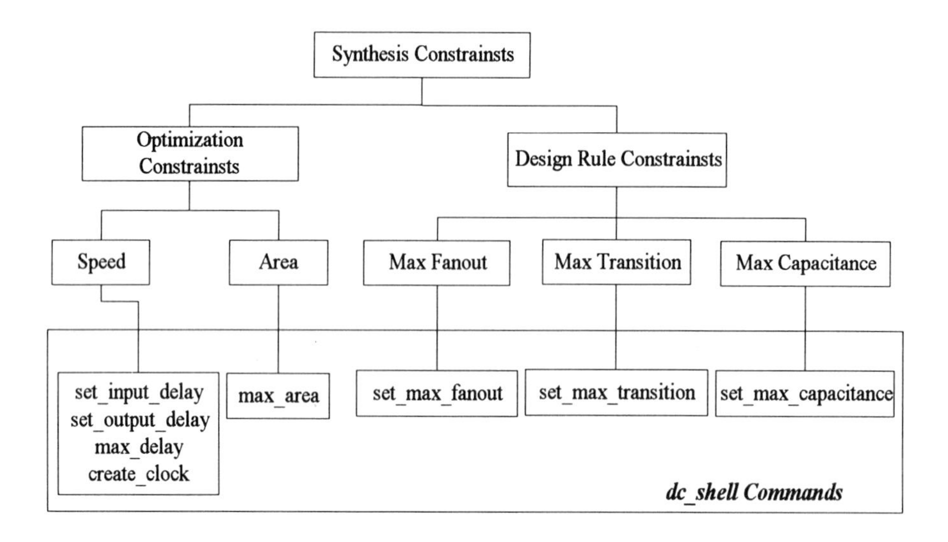


**Figure 4.1 Synthesis Constraints and Commands**

Figure 4.1 shows the two types of synthesis constraints and the related dc_shell commands. Optimization constraints are user specified constraints. The two optimization constraints are speed and area constraints. In other words, DC considers speed and area as the two criteria for optimization. In addition to optimization constraints, the synthesis tool is required to meet another set of constraints called Design Rule Constraints (DRC). DRC are constraints imposed upon the design by requirements specified in the target technology library. Thus, DRC have precedence over optimization constraints since DRCs have to be met in order to realize a functional design.

## 4.1.1 Design Rule Constraints (DRC)

The three design rule constraints are max_fanout, max_transition and max_capacitance. In this section, we introduce these terms and the related dc_shell commands to specify these constraints.

### max_fanout

It is a measure of the number of loads a pin or port can drive.

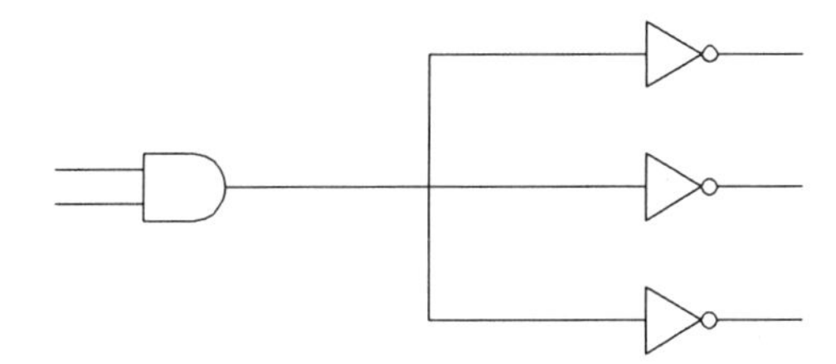

**Figure 4.2 Max_fanout**

Figure 4.2 shows the output pin of an AND gate driving the input pins of three inverters. The input pins of each of these three gates has a fanout_load attribute specified in the library. The sum of the fanout loads of each of the three input pins must not exceed the max_fanout of the output pin of the AND gate. This, in essence, is the meaning of max_fanout. It is typically an integer, although each of these fanout_load values implies a certain standard load. This is usually specific to the technology library. Input pins of library cells can have the fanout_load explicitly specified in the technology library or the library can have a global default_fanout_load for the input pins of all cells in the library. One can find the fanout load on a specific input pin of a library cell (say, AND2 in library, libA) using the following dc_shell command:

```
get_attribute find(pin, "libA/AND2/i") fanout_load
```

Instead, if the library has a default_fanout_load attribute specified in the technology library, one can find this value using the following command:

```
get_attribute libA default_fanout_load
```

## max_transition

It is the longest time for a transition from logic level 0 to 1, or vice-versa, for an entire design or for a specific net in a design. To be more specific, it is the RC time which is the product of the resistance (R) and the capacitive load (C). In DC terminology, max_transition can be defined as the product of rise/fall resistance and the capacitive load on a net. In addition to design rule constraints specified in the technology library, the user can specify these constraints as well. When the user specifies a max_transition constraint in addition to the one already specified in the technology library, the more restrictive constraint will apply. For example, if the library has a max_transition of 5 and the user were to specify a max_transition of 3, then DC will try to meet a max_transition requirement of 3.

### max_capacitance

The max_transition design rule constraint does not provide a direct control over the actual capacitance of nets. The max_capacitance design rule constraint was introduced to provide a means to limit capacitance directly. This constraint behaves like the max_transition constraint, but the cost function is based on the total capacitance of the net instead of the transition time. The max_capacitance constraint is fully independent, so one can use it in conjunction with max_transition. During compile, DC ensures that there are no max_capacitance violations, that is, the max_capacitance constraint on the output pin of a driving cell exceeds or equals the summation of the capacitance of the pins driven by this cell and the net capacitances. The max_capacitance attribute can be specified on designs or ports. max_transition, max_fanout and max_capacitance can be used to control buffering in a design.

Max_fanout, max_transition and max_capacitance constraints can be specified using the following commands:

```
set_max_transition <value> <design_name/port-name>
set_max_transition 5 test /* test is the name of a design */
set_max_fanout 5  <port_name/design_name>
set_max_capacitance 5 <port_name/design_name>
```

## 4.1.2 Optimization Constraints

Speed and area constraints are the two optimization constraints. These constraints are specified by the designer. DC assigns higher priority to timing constraints over area constraints. In other words, DC aims to meet timing constraints before performing area optimization.

Synchronous paths in the design are constrained by specifying all the clocks in the design. This is done using the create_clock command. In general, detailed timing constraints help get the best results from synthesis. Prior to version 3.0a of DC, max_delay and min_delay commands were used to specify timing constraints. But with 3.0a and subsequent versions, we recommend the usage of the set_input_delay and set_output_delay commands instead. However, for asynchronous paths use the set_max_delay and set_min_delay commands to specify point to point delays. max_delay constraints are imposed by explicit usage of the set_max_delay commands or implicitly due to clocks specified by the create_clock command. Similarly, the min_delay constraints are imposed by explicit set_min_delay commands or implicitly due to hold time requirements. However, DC fixes hold time requirements only when specified by the set_fix_hold command.

Area constraints are specified using the set_max_area command. The total area of a design is the sum of the area of all the cells used in the design and the area due to wires (if specified in the wire load model).

In general, one must ensure that the optimization script specifies constraints using the following dc_shell commands:

```
create_clock
set_input_delay
set_output_delay
set_driving_cell
set_load
max_area
```

### 4.1.3 Synthesis Cost Functions

The synthesis tool performs optimization by minimizing cost functions - one for design rule costs and the other for optimization costs. These cost values are usually displayed during optimization. The optimization cost function consists of four parts in the following order of importance:

1. Max delay cost
2. Min delay cost
3. Max power cost
4. Max area cost

The max delay cost carries the greatest priority in cost calculations. It is the sum of the products of the worst violators and the weight in each path group. Violations are said to occur when a design fails to meet the setup time requirements of sequential elements in the design. In general, all the paths constrained by a clock are grouped into one path group. Thus, each clock in the design creates a separate path group. All the remaining paths are grouped into the *default path group.* If no clocks are specified, then all paths default to a single default path group. Since the synthesis tool is primarily path based, it is possible to attach different weights to different path groups. For example, consider a design with three clocks. If each path group had the default weight (default value of the weight is 1 for all path groups), and if the worst violation in each group was 1, 2 and 3 respectively, then the max delay cost calculation is as follows:

Max_Delay Cost = (1 x 1) + (1 x 2) + (1 x 3) = 6.0

Min delay cost is second in priority after max delay in cost calculation. The min delay cost calculation is independent of path groups. It is the sum of all the worst min_delay violators. The min delay violation is calculated as the difference between the expected delay and the actual delay. A violation occurs when the expected delay is greater than the actual delay. Since min_delay is unaffected by path groups, the weightage assigned to paths has no effect on min delay calculations. For example, a design with three paths with min_delay violations of 1, 2 and 3 will have a min delay cost of:

Min_Delay = 1 + 2 + 3 = 6.0

Max power cost is only in the case of ECL technology. It is simply the difference between the current power and the max power specified. A violation implies that the former exceeds the later. Synopsys introduced *DesignPower*, a power estimation capability for CMOS technology and is due to introduce a power optimzation tool in the future. The discussion of these is beyond the scope of this book.

Max Area cost has the least priority in cost calculation. By default, the tool does not optimize for area once the timing constraints are met. In other words, if explicit area constraints are specified, DC performs area optimization. Since synthesis results are dependent to a large extent on a number of factors such as constraints, libraries and coding styles, optimization of a design is an iterative process.

# 4.2 Clock Specification for Synthesis

Clock specifications are one of the important issues for synthesis. One must define each clock in the design using the create_clock command. Clock trees are not usually synthesized using DC and hence they must be hand instantiated. It is recommended that the set_dont_touch_network command be used for clocks to prevent synthesis from buffering clock trees. This command ensures that the entire clock network in the design inherits a dont_touch attribute. In the event of a hand instantiated clock tree, during synthesis, one must also place a dont_touch attribute on the clock network using the set_dont_touch_network command.

DC considers all clock network delays to be ideal. In other words, the Synopsys static timing analyzer (DesignTime) does not report timing delays to the clock pin of a sequential element. If the design has a gated clock, the tool by default, will not consider the delay through the gates leading to the clock. One can override this default behavior by using set_clock_skew command with the -propagated option to obtain non-zero clock network delay. If your ASIC vendor has specified an upper limit of potential skew on your chip between clock pins of sequential elements then use the set_clock_skew -uncertainty command to specify the skew. The recommended methodology for specifying the timing specifications in DC is to specify the timing with respect to the clocks and the clock edges. Hence, after specifying the clocks using the create_clock command, specify all the input and output port timing specifications using the set_input_delay and the set_output_delay command with the -clock option. Further, Synopsys, by default, assumes single cycle paths. To specify multi-cycle paths use the set_multicycle_path command.

## 4.3 Design Compiler Timing Reports

DC reports timing delays from clocks-to-clocks. In other words, DC reports timing from synchronous logic to synchronous logic or the logic between sequential cells. By default, timing paths end at pins which have a setup/hold constraint like the data pin of a flip flop and asynchronous pins of registers (set, clear). This behavior can be modified for asynchronous pins of registers by executing the report_timing command with the enable_preset_clear option.

The DC timing report, by default, lists only the worst path in each path group. Each clock declared by the create_clock command creates a separate path group. The report_timing command shows the path from primary input/clock pin to the primary output/data pin unless startpoints/endpoints have been created at pins internal to the design by the explicit usage of set_input_delay and set_output_delay commands. In this section, we consider a simple example to explain timing reports generated by DC. The effect of timing constraints such as set_input_delay and set_output_delay on DC timing reports is also discussed.

### Example 4.1 Design Showing Flop to Flop Delays

```
module timing(a, b, clk,reset, d) ;
input a, b, clk , reset ;
output d;
reg a, b ;
reg f , d;

always@ (posedge  clk or negedge reset)
begin
   if (!reset)
      f <= 0 ;
   else
      f <= a ;
end
always@ (posedge  clk or negedge reset)
begin
   if (!reset)
      d <= 0 ;
   else
      d <= f & b ;
 end
endmodule
```

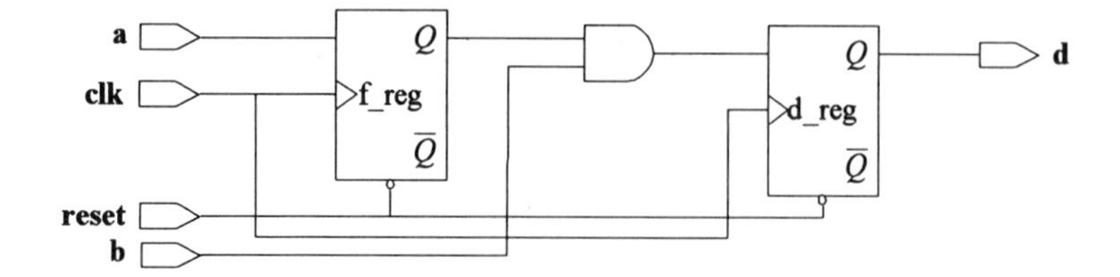


**Figure 4.3 Generated Schematic**

Example 4.1 shows the Verilog code and the schematic generated (Figure 4.3). The dc_shell script used to compile the above design is as follows:

```
read -f verilog test.v
link_library = target_library = lsi_10k.db
create_clock clk -period 5
compile
/* The following command generates the timing report shown below */
report_timing -max_paths 5
****************************************

Report : timing
        -path full
        -delay max
        -max_paths 5
Design : timing
****************************************

Operating Conditions:
Wire Loading Model Mode: top
  Startpoint: f_reg (rising edge-triggered flip-flop clocked by clk)
  Endpoint: d_reg (rising edge-triggered flip-flop clocked by clk)
  Constraint Group: clk
  Path Type: max
  Point                          Incr     Path
  -----------------------------------------------
  clock clk (rise edge)          0.00     0.00
  clock network delay (ideal)    0.00     0.00
  f_reg/CP (FD2)                 0.00     0.00 r
  f_reg/Q (FD2)                  1.42     1.42 f
  U33/Z (AN2)                    0.82     2.24 f
```

```
d_reg/D (FD2)                     0.00     2.24 f
data arrival time                            2.24
clock clk (rise edge)             5.00     5.00
clock network delay (ideal)       0.00     5.00
d_reg/CP (FD2)                    0.00     5.00 r
library setup time               -0.85     4.15
data required time                          4.15
-------------------------------------------------------
data required time                         4.15
data arrival time                         -2.24
-------------------------------------------------------
slack (MET)                                1.91

Startpoint: a (input port)
Endpoint: f_reg (rising edge-triggered flip-flop clocked by clk)
Constraint Group: clk
Path Type: max

Point                                  Incr       Path
---------------------------------------------------------------
clock (input port clock) (rise edge)   0.00      0.00
input external delay                   0.00      0.00 r
a (in)                                 0.00      0.00 r
f_reg/D (FD2)                          0.00      0.00 r
data arrival time                      0.00

clock clk (rise edge)                  5.00      5.00
clock network delay (ideal)            0.00      5.00
f_reg/CP (FD2)                         0.00      5.00 r
library setup time                    -0.85      4.15
data required time                                4.15
---------------------------------------------------------------
data required time                              4.15
data arrival time                               0.00
---------------------------------------------------------------
slack (MET)                                        4.15
```

Let us analyze these timing reports. The report gives the point in the design, which is usually a port or a pin of a library cell, the incremental delay through the cell (listed in the Incr column), and the Path delay (listed under the Path column) or the delay in the path up to that point. In other words, the path delays are calculated by adding up the incremental delays.

Consider the first path beginning at the first sequential element f_reg and ending at the next sequential element d_reg in Figure 4.3. The rising edges of the clock are at 0 and 5 ns. So for f_reg, assuming no clock network delays (which is the default condition), the clock rise occurs at 0 ns, the clock to Q delay of the FD2 flop is 1.42 ns, the delay through AND gate is 0.82 ns, giving a data arrival time of 2.24 ns at the data pin of d_reg. The register d_reg has its first rising edge at 0. At this stage data from f_reg had not yet arrived. However, for the next rising edge at 5 ns, the situation is different since data from f_reg arrived at 2.24 ns. Now, since the rising edge is at 5ns and the library has a setup requirement of 0.85 ns for FD2 flop, the latest a signal can arrive to avoid setup time violations is 5 - 0.85 = 4.15 ns. This implies that the constraint has been met with a positive slack of 1.91 ns.

Notice that by specifying a clock period of 5 ns, we have implicitly placed a max_delay constraint of 4.15 ns from clock pin of f_reg to data pin of d_reg. In other words, the setup time requirements imposed by the technology library and the clock period specified by the create_clock command places an implicit max_delay constraint on paths between synchronous elements. In the second report, data arrives at 0 ns, and the clock requires that data arrive latest by 4.15 ns implying that setup time is met with a slack of 4.15 ns.

### Fix_hold

By default, DC does not fix hold time violations. To ensure that DC inserts delays to meet hold time requirements one must use the set_fix_hold command as shown below.

```
fix_hold clk
```

## 4.3.1 Report After Setting an Output Delay

The timing report after setting an output delay of 2 ns on output port d using the set_output_delay command is shown below. The output delay specified on port d implies that the path must account for a delay of 2 ns, external to the Q pin of d_reg. In other words, data must arrive 2 ns earlier at the output port d in order to meet timing requirements defined by flip-flops external to the port d.

```
****************************************
Report : timing
        -path full
        -delay max
        -max_paths 1
Design : timing
****************************************
Operating Conditions:
Wire Loading Model Mode: top

  Startpoint: d_reg (rising edge-triggered flip-flop clocked by clk)
  Endpoint: d (output port clocked by clk)
  Constraint Group: clk
  Path Type: max

  Point                                  Incr       Path
  -----------------------------------------------------------
  clock clk (rise edge)                  0.00       0.00
  clock network delay (ideal)            0.00       0.00
  d_reg/CP (FD2)                         0.00       0.00 r
  d_reg/Q (FD2)                          1.37       1.37 f
  d (out)                                0.00       1.37 f
  data arrival time                      1.37

  clock clk (rise edge)                  5.00       5.00
  clock network delay (ideal)            0.00       5.00
  output external delay                 -2.00       3.00
  data required time                     3.00
  -----------------------------------------------------------
  data required time                                 3.00
  data arrival time                                 -1.37
  -----------------------------------------------------------
  slack (MET)                                        1.63
```

Note the output external delay specified in the timing report.

### 4.3.2 Timing Report after setting an input delay constraint

The following report was generated after specifying an input delay of 3 ns on the input port A using the set_input_delay command. set_input_delay is similar to set_output_delay, except that it accounts for timing delays at the input. For example, in the following report, an input_delay of 3 ns on input port A, implies that relative to the rising edge of clock, clk, there is a delay of 3 ns, due to logic or otherwise prior to the port A.

```
****************************************
Report : timing
        -path full
        -delay max
        -max_paths 5
Design : timing
****************************************

Operating Conditions:
Wire Loading Model Mode: top
  Startpoint: a (input port clocked by clk)
  Endpoint: f_reg (rising edge-triggered flip-flop clocked by clk)
  Constraint Group: clk
  Path Type: max
  Point                                   Incr      Path
  ----------------------------------------------------------------
  clock clk (rise edge)                   0.00      0.00
  clock network delay (ideal)             0.00      0.00
  input external delay                    3.00      3.00 r
  a (in)                                  0.00      3.00 r
  f_reg/D (FD2)                           0.00      3.00 r
  data arrival time                       3.00

  clock clk (rise edge)                   5.00      5.00
  clock network delay (ideal)             0.00      5.00
  f_reg/CP (FD2)                          0.00      5.00 r
  library setup time                     -0.85      4.15
  data required time                      4.15
  ----------------------------------------------------------------
  data required time                      4.15
  data arrival time                                   -3.00
  ----------------------------------------------------------------
  slack (MET)                                          1.15
```

## 4.4 Commonly Used Design Compiler Commands

In this section, we discuss a few basic DC commands and switches and their usage. While this is only an introductory list, these are fundamental to understanding optimization strategies for DC.

### 4.4.1 set_dont_touch

This is a very useful command, particularly when optimizing hierarchical designs. It is an attribute which is assigned to a design or library cell. After one has specified the constraints and compiled a design to meet the constraints, it is often required that this design not be re-optimized when used in a larger design. In such cases, one would specify a dont_touch attribute on the instance of that design in the higher level design. For example, say block A has been optimized to satisfaction and has been used in another design TOP as shown in Figure 4.4. Say the instance name of block A in TOP is u1, then the following dc_shell script performs the dont_touch step:

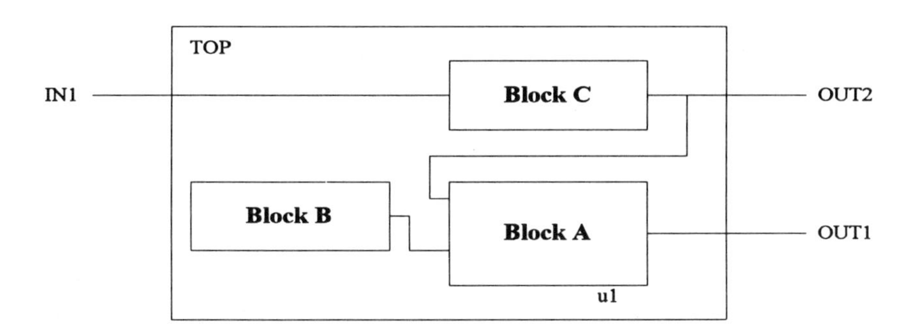


**Figure 4.4 dont_touch in Design with Hierarchical Blocks**

```
current_design = TOP
set_dont_touch u1 /*or alternatively, set_dont_touch find(cell, u1) */
```

The dont_touch attribute when placed on a reference is inherited by all instances of the reference in the design. For example, if one were to place a dont_touch on Block A in Figure 4.4 as shown below:

```
current_design = BlockB
set_dont_touch find(design, BlockA)
```

Then, every instance of Block A will inherit the dont_touch attribute. In this example, instance u1 inherits the dont_touch attribute. Both the report_reference or the report_cell commands will show this attribute. If the dont_touch attribute is placed on the reference, one will have to first remove this attribute from the reference. In other words, it is not possible to remove the dont_touch attribute from an instance whose reference has the dont_touch attribute. The dont_touch attribute can be removed using the remove_attribute command.

```
remove_attribute find(design, A) dont_touch
remove_attribute find(cell, u1) dont_touch
```

To find the cells in your design which have the dont_touch attribute, use the filter and find commands. To search the entire hierarchy, use the find command with the -hierarchy option. Appendix A has several examples using these commands.

```
filter find(cell -hierarchy, "*") "@dont_touch == true"
```

Another use of this command is when instantiating a library cell. If one wishes to ensure that this instantiated cell remain in the netlist after optimization, one should place a dont_touch on the instance of the library cell prior to the compile step.

### 4.4.2 Flattening and Structuring

Flattening a design essentially means converting the combinational logic into a two-level sum of products form. This is usually done to improve the speed of the design. However, flattening has its limitations and is not recommended for designs or blocks which contain XORs, multiplexors and adders. Flattening is not performed by default during the optimization process. Intermediate terms are removed when flattening a design. For a design with over 20 inputs, flattening is almost never completed by DC. If the number of inputs is less than ten, flattening is more likely to be completed. Flattening does not complete since most libraries usually do not have cells with more than 5 inputs (5 input AND gates, for example) which are required to achieve a two level sum of products form. In the event of flattening not being completed by DC, it simply proceeds to the next step after issuing a message that "Flattening is too expensive...".

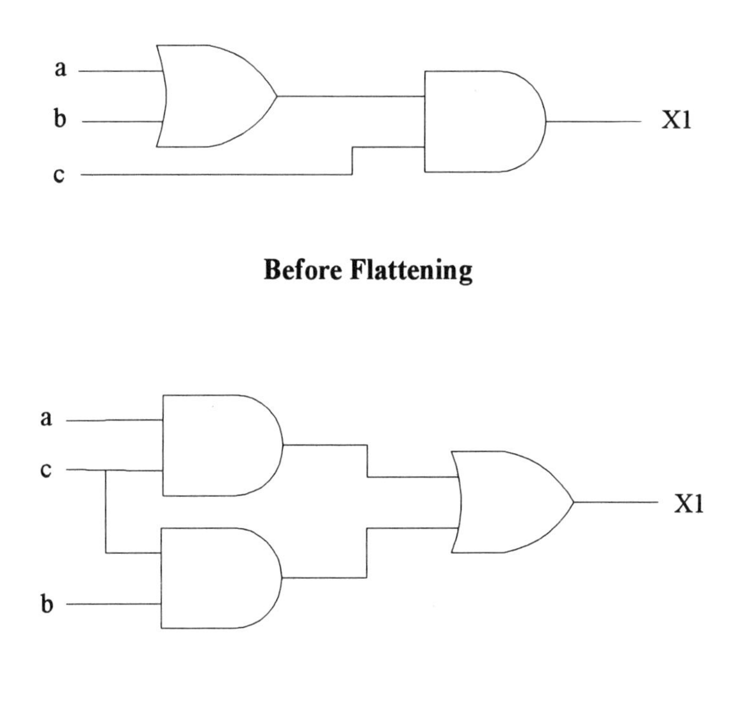

**Figure 4.5 Gate Structures Before and After Flattening**

### Example 4.2 Boolean Equation Before and After Flattening

Y1 = a + b ;
X1 = Y1. c ;

Flattening the above equations results in:

X1 = ac + bc ;

Example 4.2 shows boolean equations before and after synthesis with the flatten switch in DC turned on. (Figure 4.5). Structuring on the other hand, is used to improve the area or gate count of a design. It involves the addition of intermediate terms which are then shared by different outputs. In this sense, it can be considered as a reverse process of flattening.

Structuring is of two kinds - *timing driven* and *boolean structuring*. While timing driven structuring is executed by default, the latter is not. Timing driven structuring takes into account time delays when structuring the design, while boolean structuring does not. Further boolean structuring results in a 2X to 4X increase in compile time. This can be very significant when dealing with large designs. Hence, boolean structuring is not often recommended. Boolean structuring can be applied to random logic in order to minimize area. Example 4.3 shows the boolean equations illustrating the concept of structuring a design.

### Example 4.3 Boolean Equation Before and After Structuring

X1 = ab + ad ;

X2 = bc + cd ;

After structuring,

Y1 = b + d ;

X1 = a. Y1 ;

X2 = c. Y2;

Flattening and structuring can be specified using the following DC commands.

```
set_flatten true
set_structure -timing true
```

To verify these options use the following command:

```
report_compile_options
```

### Flattening vs. Ungrouping

Flattening is often mistaken for *flattening the hierarchy* of a design. Flattening the hierarchy in DC is referred to as *ungrouping the design* and is independent of the *flatten* compile option. Flatten is used throughout this book in the context of compile option, and ungroup is used to mean removing the levels of hierarchy in a design. Thus, a flattened design does not mean a design devoid of hierarchy, but implies a design optimized with the flatten option turned on.

## 4.4.3 Ungroup and Group

The ungroup and group commands are used to remove and create levels of hierarchy in a design respectively. The ungroup command, when executed, causes the instances in the previously existing level of hierarchy to inherit the names of the level which was ungrouped. For example, if a sub-block A in hierarchical design TOP is ungrouped, an instance u2 previously contained in A will have the new instance name a1/u2, where a1 is the instance name of the reference sub-block A.

The same naming convention applies to nets in the design ungrouped. This helps particularly when one wishes to re-group cells into a new level of hierarchy or for back annotation purposes. However, it is possible to override this type of naming convention by executing the ungroup command with the simple_names option. One can recursively ungroup the different levels in a hierarchy by executing the ungroup command with the flatten option. For example, with current_design set to the top level of the design, if one were to execute the following command:

```
ungroup -flatten -all
```

then all levels below top are recursively ungrouped, provided none of the hierarchical blocks have the dont_touch attribute applied to them. The replace_synthetic -ungroup command can be used to ungroup all the synthetic designs before compiling the design.

The group command is the inverse operation of ungroup. One can group several instances in a design into a new level of hierarchy and specify the design name (the reference name) and the instance name of this design in the level of hierarchy above this newly created instance. For example, using the following command one can group instances u1 and u2 in a design TOP, to form a new design new_block within TOP and then assign the instance name Z1 to it.

```
group{u1 u2} -design_name new_block -cell_name Z1
```

The group command can also be used to group logic associated with block/process statements in the source Verilog/VHDL. In this case, one will have to execute the group command with hdl_block option.

### 4.4.4 set_dont_use

When one wishes to prevent DC from inferring certain cells in the technology library, the set_dont_use command must be used. It assigns the dont_use attribute to the library cells specified. The cells with the dont_use attribute are not used or ignored during optimization. For example, one can place a dont_use attribute on the library cell NAND2 in library libB as shown below:

```
set_dont_use libB/NAND2
```

### 4.4.5 set_prefer

The set_prefer command changes the priority of cells chosen by DC during technology translation. Technology translation is essentially the process of mapping a netlist from one technology library to another. This does not, however, have any effect when executing the compile command. This command assigns the prefer attribute to the specified cells. For example, one can place a prefer attribute on the library cell IVB in library libB. This causes DC to infer the IVB cell each time a cell of that functionality is required.

```
set_prefer libB/IVB
```

### 4.4.6 characterize

The characterize command is used extensively in hierarchical designs. For example, consider a design TOP with two sub-blocks sub1 and sub2 as shown in Figure 4.6. Let us assume that both sub1 and sub2 have been compiled individually and have met their constraints. However, when instantiated in TOP, sub1 and sub2 have different constraints depending on constraints on TOP and the logic surrounding sub1 and sub2 in TOP. Hence, these new constraints might not necessarily be met.

The characterize command helps capture the constraints imposed on the sub-design by the surrounding logic. Consider the dc_shell script shown below. The write_script command writes the constraints to a file. It is advisable to characterize and compile each instance separately, since after the characterize-compile of one sub-block, the characterize data for the other block is quite different.

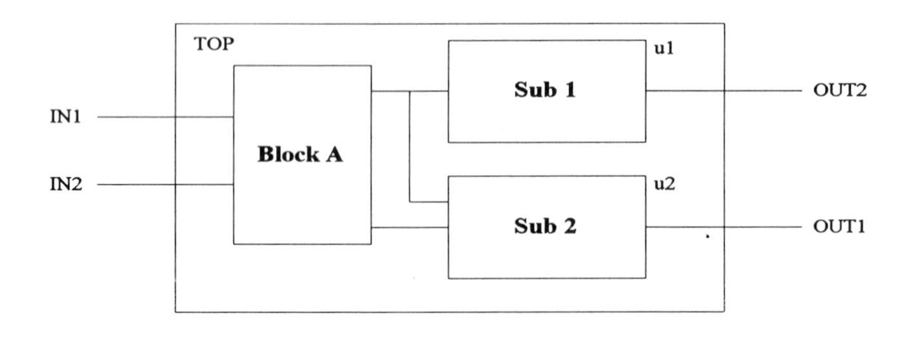


**Figure 4.6 Characterizing Hierarchical Blocks**

```
read -f db TOP.db
/* specify constraints from TOP */
characterize u1 /* instead of characterize {u1 u2} */
/*u1 is the instance name for sub1, say */
current_design = sub1
write_script > sub1.scr
compile
current_design TOP
```

```
characterize u2
current_design = sub2
write_script > sub2.scr
compile
```

## 4.5 Strategies for Compiling Designs

What is the best approach to compiling a design? Unfortunately, there is no simple answer to this question. The compile strategy adopted is very much design dependent. Further, synthesis results are a function of the technology library, the coding style and the synthesis strategy. However, it is possible to follow some general guidelines for compiling a design. But, first we discuss the compile *approaches that must be avoided.*

1. Capturing the entire design in one large HDL file, reading that file into DC, specifying the following constraint:

   ```
   max_delay 0 -from all_inputs() -to all_outputs()
   ```

   followed by executing the miraculous `compile`. There can be no greater formula for disaster than this approach. The current state of synthesis technology is far from push-button.

2. Dividing your design into too many hierarchical sub blocks. This is the other extreme of strategy 1. This is not recommended for two reasons. Firstly, managing the design with several sub-blocks can be rather cumbersome. Secondly, optimization across hierarchical boundaries is not as effective as optimization within a block. But again, if this is one huge block of logic, DC is not fully effective in optimization (as in case 1). In general, a recommended size of a module for optimization is in the range 250-5000 gates. Again, there could be exceptions to this rule.

From the above, clearly, one has to strike a *middle-of-the-road* approach. This approach is very much design dependent and can be arrived at only through several iterative steps.

## 4.6 Typical Scenarios When Optimizing Designs

In this section, we discuss all the different strategies that one can experiment with when optimizing a design using DC. Here we assume that we are dealing with just one sub-block of design and not the entire design. We attempt to address the fundamental issue, "I have one single sub-block, what is the best way to go about optimizing it?"

**Scenario 1**

You have a design written in HDL. You have a very limited idea of the timing requirements. You simply wish to attain the fastest possible design.

A simple strategy to realize the optimal design is to experiment first with a default, medium effort compile, specifying absolutely no constraints before the compile step. This should give you a feel for the timing/area performance of your block. Then, specify your approximate timing (clocks and point to point timing constraints, if any) requirements. Then, analyze your results using the following dc_shell command:

```
report_constraints -all_violators -verbose
```

DC lists all the paths that fail to meet timing requirements in the design. If a number of paths are violated by a large margin, you know right away that meeting your timing is likely to be a difficult/impossible task. Then, re-compile the design after reading in the source code and specifying optimization constraints. On the other hand, if very few paths violate timing, re-compile and analyze the results. DC works best when you set achievable targets or, rather, go about achieving your goals in a step by step process. This is also an outcome of the fact that synthesis results depend largely on the starting point of your compile step. In general, when you are unable to meet timing, one or more of the following approaches can be followed.

- Re-assess your code and consider alternate design partitioning.
- Modify your constraints to set more realistic constraints.
- Identify any functional false paths or multi-cycle paths that might exist and specify them.
- Specify point to point max_delay/min_delay timing constraints for asynchronous paths.
- For synchronous paths that violate timing use the group_path command to create separate path groups and assign weights to different path groups.
- If you are very close to meeting timing at this stage, try an incremental compile.

Most often you will find that your synthesis steps, by and large, follow a pattern similar to the approach explained above. In general, the results are always a function of the starting point of the design. Depending on the starting point, the end result after compile can be drastically different. A sound knowledge of the capabilities of the cells in your technology library is extremely helpful. It is also possible that the constraints you are trying to meet can't be achieved using the cells currently available in the technology library.

**Scenario 2**

You have written your source code, you know the detailed timing requirements, from characterize or otherwise. You wish to meet these timing requirements.

Execute the following steps in DC using the appropriate commands:

```
Read in your design
Specify timing/area constraints
Synthesize the design using the compile command
Analyze results using report_constraints -all_violators -verbose
```

If design does not meet constraints, use the group_path command to assign higher weightage to paths which show greater violations. The compile_default_critical_range variable can be used to ensure that DC not only works on the worst violator. For example, when this variable is set to the value 2 DC looks at all paths within 2 ns of possible violation of timing. The final step could be an incremental compile. This is used only to make very minimal improvements in timing, usually less that 2 ns. In general, the more specific you can be in specifying constraints, the better the synthesis results. In short, execute the following steps before re-compiling the design.

```
Identify any multi-cycle paths and false paths that might exist in the design
Create path groups using group_path command.
Set the variable compile_default_critical_range
compile
compile incremental -map_effort high
```

If you are still unable to meet timing, begin with the source HDL and re-compile using updated synthesis script that identifies multi-cycle paths and path groups.

**Scenario 3**

You have fairly accurate timing requirements, but your main motive is to improve rather than merely meet the requirements. You are confident from knowledge of your library cells and earlier compile iterations that DC can meet timing, but your intent is to get the fastest possible design.

```
Read in design
compile
Specify constraints
report_timing
If constraints are already close to being met, specify tighter constraints.
compile
```

You now meet timing but wish to improve upon this.

```
report_constraints -all_violators -verbose
```

Now specify tighter constraints -- faster clock and tighter max_delay constraints for asynchronous paths. Execute report_constraints, report_timing and analyze your results. Do not specify unrealistic constraints, like max_delay 0 for instance. Instead, gradually tighten constraints. If the current timing is 8 ns. First try, say 6.5, if that is met (rather easily, in short compile time), then be more aggressive, say, 5.0 and so on.

In summary, to optimize for timing, if you know your constraints in detail, apply them up front after reading in your source HDL. On the other hand, if you do not have detailed constraints for your design (which is most likely the case!), execute an initial compile to get a feel for what DC gives you. Analyze the results with regard to slack and the library cells used. Then, apply constraints and re-compile. The drawback of an initial compile with no constraints is that the structure of the design inferred is more or less fixed after this compile, thereby limiting the possibility of other more efficient structures. Hence, once you have greater detail of constraints, it is advisable to revert back to the earlier strategy of applying constraints up front to compare results. After every compile step, assess your results. If the results are the best you have received so far, save the design in db format.

**Scenario 4**

Area is extremely critical in your design. While you think you could meet timing, area is an issue you would like to monitor right from the very start of your synthesis process. Given below are some tips for effective area optimization:

- Prior to the initial compile one must try and specify very accurate constraints to prevent DC from overkill of non-critical paths.
- After synthesis, execute the check_design command. Analyze the results to make sure there is no unused logic in the design. Useful details about the design such as unconnected ports, feedthroughs, and multiple drivers are provided by this command.
- Use the report_resources command to check implementations of resources in the designs and also on how many resources are inferred. There might be scope for sharing of resources by modifying the HDL code.
- Try ungrouping the hierarchy. Although this might improve area, it might make place and route task extremely difficult.

- Flatten appropriate unstructured random logic blocks using the set_flatten command on these blocks. Group control logic and datapath logic into separate designs using the group -hdl_block command, so that different compile options can be used.

In short, synthesizing a design is an iterative process. The process can be refined by extensive usage of the tool, analysis of the results, and a fair knowledge of the capabilities of the cells in the technology library.

### 4.6.1 Compiling Hierarchical Designs

In most cases, hierarchical designs are not compiled top down or bottom up. It is usually a combination of the two. This is most common when a chip uses existing sub modules available either in HDL netlist form or in schematics. Often, the design is a combination of pre-existing design blocks and newly designed blocks. In such cases, from the top level one has a fair idea of the constraints. In such a scenario, your approach should be as follows:

1. Read in all the sub blocks of your design, and finally your top level.
2. Specify constraints at the top level of the design.
3. Characterize and compile each sub-block individually. In other words, you have two sub-designs s1 and s2 (instance names are u1 and u2 respectively) contained in top. The dc_shell script will be as follows:

```
current_design = top
characterize u1
current_design s1
compile
current_design = top
characterize u2
current_design = s2
compile
```

The above approach is the most commonly used synthesis strategy and is called the *compile-characterize-compile* approach. In other words, the sub-blocks are compiled individually using the strategy explained in Scenario 1 to get a fair estimate of the timing capabilities of the sub blocks. Then, the different sub-blocks are characterized and re-compiled. If you have multiple clocks at the top level, completion of the characterize step can take rather unacceptable time delays and sometimes run out of memory. Such a situation can be countered using one of two approaches. Partition your design such that you can specify timing constraints for blocks below the top level. Then, characterize at the level below the top level. Try to ensure that you do not have more than two clocks (preferably one) driving the same sub-block.

An alternative is to try a combination of approaches. Compile some of your lower blocks first with approximate timing constraints. In other words, you follow a *time budgeting* approach. Based on your understanding of the constraints at the top level you budget timing delays in the different lower blocks. This is usually dependent on a very sound understanding of the design's timing requirements and is the most difficult of the compile strategies.

## 4.7 Guidelines for Logic Synthesis

The guidelines suggested here are not "hard and fast" rules for effective synthesis. These are applicable in most cases and exceptions to these guidelines are possible.

1. For better results from synthesis, specify accurate point to point delays for asynchronous paths. Use the create_clock and group_path commands to constrain synchronous paths in the design. In general, the synthesis tool is tailored towards path optimization. Hence, it responds better to a greater detail of constraints.
2. Try to register outputs of the different design modules. This saves the designer from having to perform painstaking time budgeting. Constraining different hierarchical modules becomes easier for two reasons. The drive strength on the inputs to a block is equal to the drive strength of the average flip flop. Secondly, the input delays are equal to the path delays through a flip flop, given that the outputs of the driving hierarchical block are registered.
3. Separate negative and positive edge flip-flops into separate hierarchical blocks. In other words, avoid having both kinds of flops in the same hierarchical module. This makes the debug process and timing analysis during synthesis much simpler. Moreover, this can help simplify test insertion.
4. Group finite state machines and optimize them separately. State machine extraction and optimization process is more effective when the fsm is isolated. The group -fsm command can be used to achieve this.
5. The recommended size of a module for synthesis is in the range 250-5000. There are bound to be exceptions to this generalized recommendation.
6. Avoid having too many hierarchical blocks. Optimization across hierarchical boundaries is far less effective than when the boundaries do not exist. On the other hand having a large flat design with no hierarchy is not the solution. One has to develop a feel for the "middle-of-the-road" strategy.
7. Try to capture logic in the critical path into a separate level of hierarchy. DC does a better job of optimization when the critical path does not traverse hierarchical boundaries. This can be done by ungrouping existing blocks and re-grouping them using dc_shell scripts. Similarly, when declaring point to point false paths,

it is preferable that the path lies in a single hierarchical module. When they traverse different hierarchical modules, they could significantly increase compile time.

8. If your compile time is too long, it is most likely due to one of the following reasons:

   – You are using high map effort. Try the default medium effort. This is the recommended compile effort and hence is the default. The compile time for high map effort is dependent on the machine configuration and the size of the design.

   – Your design is too large and must be broken down into smaller hierarchical modules.

   – You have declared false paths which traverse hierarchical boundaries or any path exceptions specified in the design such as **set_multicycle** paths.

   – You have glue logic at the top level of your design. Consider incorporating this into hierarchical sub modules using the **ungroup/group** commands.

   – You are trying to flatten a design which is not appropriate for flattening. In general, use the **flatten** switch only for random logic. Blocks containing logic such as adders muxes and XORs are not appropriate for flattening. For a design with over twenty inputs, flattening is almost never completed. If the number of inputs is less than ten, flattening is more likely to complete.

   – You have boolean optimization turned on. Again, this is appropriate only for random logic. If you do have random logic in your design, consider grouping it into a separate level of hierarchy and compile it separately with the flatten or boolean structuring switch turned on.

   – Last but not the least, consider adding memory to your machine!

9. For datapath logic, consider the option of instantiating logic (like gates and muxes) or inferring them through user developed DesignWare libraries as discussed in Chapter 10. It is possible that you will achieve a better timing than what DC can infer from discrete gates.

10. Partitioning the design is extremely crucial to get the best out of synthesis. Bear in mind that compile strategies differ from module to module depending on whether it is a datapath module, or random logic or well-structured logic. Identify signals with large fanouts and attempt to group the driving logic with the logic being driven into one hierarchical block. Similarly, to facilitate sharing of resources like adders, ensure that relevant resources are in the same level of hierarchy.

11. The compile_default_critical_range = 2.0 variable ensures that DC not only works on the worst violator but also looks at all paths within 2 ns of possible violation of timing. This value of 2 is only an example and the appropriate value for this variable is usually design dependent. This variable is usually set when you are close to achieving timing, that is, +/- 2.0ns.
12. Last but not the least, it is always advisable to perform a preliminary synthesis and place and route so as to identify any serious issues which may require re-writing the HDL code.

## 4.8 Classic Scenarios

**Scenario 1**

You wish to find all the clocks defined in your design and their clock periods within a dc_shell script file. Using this information, you then wish to specify some constraints and attributes related to the clocks.

**Solution**

This can be done using the following commands:

```
find(clock, "*")
/* This should find all the clock objects created in the design */
get_attribute find(clock, clk) period
/*This should find the value of the attribute "period" for the clock object "clk" */
```

**Scenario 2**

On synthesizing a mux structure using the concurrent selected signal assignment statements you find that one of the conditions of the primary inputs is not used (that is, left unconnected). Why?

**Solution**

Consider the following code which describes a similar scenario. Signal rx_sel is a concatenation of five primary inputs. RX2 is unconnected after synthesis.

```
rx_sel <= ( RX1 & RX2 & RX3 & RX4 & RX5 );
with rx_sel select
  R0REGRD(15 downto 0)
        <= "0000111100001111"  when "10000",
          "1010101010101010"   when "01000",
          "0000000000000000"   when "00100",
          "1111111111111111"  when "00010",
```

```
        "1111100001111000"     when "00001",
        "----------------"     when others ;
```

Notice that for the unspecified combinations, the output is a dont_care. In other words, the output can take any value for unspecified combinations. So DC, in order to optimize logic, assumes the same output for both the others case and when "01000". In other words,

```
ROREGRD(15 downto 0) <= "1010101010101010" when "01000",
    "1010101010101010" when "00000",
```

Hence, the logic is independent of RX2. Synthesis chooses the same output, "1010101010101010" for all unspecified conditions and the condition "01000".

**Scenario 3**

Can one specify dont_care conditions for the condition branches of a case statement?

**Solution**

A typical scenario is when one cares only about certain inputs in a particular state but not the other inputs. DC does not support dont_cares for case statement conditions because of simulation mismatches. In the simulation world, a string to string matching is performed and this applies to the dont_care conditions as well.

**Scenario 4**

A design has a latch whose output is connected to the input of another flip-flop through muxes. DC reports the following violation:

```
Startpoint: rct_reg/ltch_6/Q_reg
  (positive level-sensitive latch clocked by CLK')
Endpoint: rctc_reg/dffwen_6/Q_reg
        (rising edge-triggered flip-flop clocked by CLK)
Path Group: CLK
Path Type: max

Point                                   Incr      Path
-----------------------------------------------------------------
clock CLK' (rise edge)                  12.50     12.50
clock network delay (ideal)              0.00     12.50
time given to startpoint                11.70     24.20
rct_reg/ltch_6/Q_reg/d (ld00p1c)         0.00     24.20 f
rct_reg/ltch_6/Q_reg/q_n (ld00p1c)       1.47     25.67 r
```

```
U444/z (ao222)                             0.92    26.59 r
U445/z (ao22I)                             1.32    27.91 r
rctc_reg/dffwen_6/Q_reg/d (fd00p1c)        0.00    27.91 r
data arrival time                                  27.91

clock CLK (rise edge)                     25.00    25.00
clock network delay (ideal)                0.00    25.00
rctc_reg/dffwen_6/Q_reg/c (fd00p1c)        0.00    25.00 r
library setup time                        -0.69    24.32
data required time                                 24.32
----------------------------------------------------------------
data required time                                 24.32
data arrival time                                 -27.91
----------------------------------------------------------------
slack (VIOLATED)                                   -3.60
```

Why is the timing report showing "time given to startpoint" and why is there a violation?

**Solution**

The period of the clock CLK is 25 with a rising edge at 0 and falling edge at 12.5. The latch enable is driven by the inverted CLK. For a positive level sensitive latch, the maximum time that can be borrowed is the duration of the active clock pulse - the setup time of the latch. In this case the timing path terminating at the latch rct_reg/ltch_6/Q_reg has borrowed 11.70 ns to meet the setup requirements at the latch. Hence, in the timing report of the path starting at rct_reg/ltch_6/Q_reg and terminating at rctc_reg/dffwen_6/Q_reg, the data startpoint is 12.5 + 11.70 rather than 12.5 (where 12.5 is the time when the latch gets enabled and 11.70 is the time at which data becomes valid). Note that even though time borrowing has occurred in the previous path, it might still be violated. Use the set_max_time_borrow command to control the time borrowing of latches.

**Scenario 5**

A design has a latch whose input is connected to a bidirectional tri-state bus as shown in Figure 4-7. The latch output is connected to the same bidirectional bus through a tri-state buffer. The write enable and read enable are mutually exclusive.

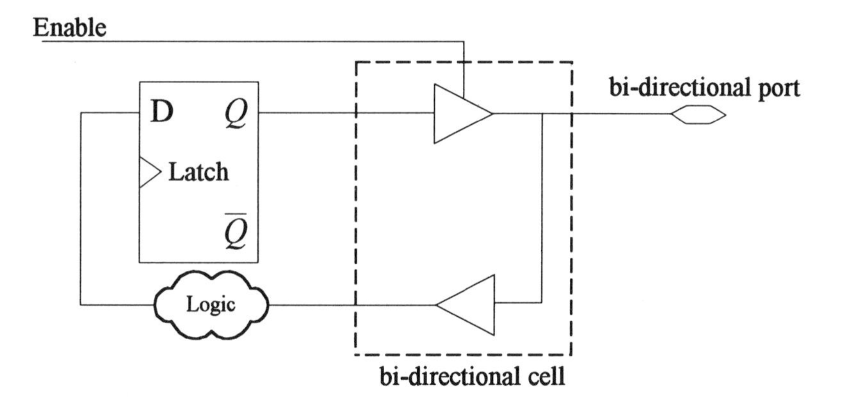


**Figure 4.7 Latch with tri-state bus at the output**

Why does DC generates the following violation? In other words, why does DC report timing from the latch to the tri-state buffer, and back to the input of the latch, although the read and write enable lines are functionally exclusive?

```
Performing report_timing on pin 'temp_reg/D'.
****************************************
Report : timing
       -path full
       -delay max
       -max_paths 1
Design : bidir
****************************************
Operating Conditions:
Wire Loading Model Mode: top

  Startpoint: temp_reg (negative level-sensitive latch clocked by clk)
  Endpoint: temp_reg (negative level-sensitive latch clocked by clk)
  Path Group: clk
  Path Type: max

  Point                                         Incr      Path
  ---------------------------------------------------------------
  clock clk (fall edge)                        12.50     12.50
  clock network delay (ideal)                   0.00     12.50
  time given to startpoint                     12.10     24.60
```

```
temp_reg/D (LD2)                         0.00    24.60 f
temp_reg/Q (LD2)                         1.53    26.13 f
U32/Z (BTS4)                             0.87    26.99 f
U31/Z (IVP)                              0.43    27.42 r
U30/Z (EON1)                             0.95    28.37 f
temp_reg/D (LD2)                         0.00    28.37 f
data arrival time                                28.37

clock clk (fall edge)                   12.50    12.50
clock network delay (ideal)              0.00    12.50
temp_reg/GN (LD2)                        0.00    12.50 f
time borrowed from endpoint             12.10    24.60
data required time                               24.60
------------------------------------------------------------------------
data required time                               24.60
data arrival time                               -28.37
------------------------------------------------------------------------
slack (VIOLATED)                                  3.77

Time Borrowing Information
------------------------------------------------
clk pulse width                        12.50
library setup time                     -0.40
------------------------------------------------
max time borrow                        12.10
actual time borrow                     12.10
------------------------------------------------
```

**Solution**

DC analyzes any timing path by breaking up the path into a head-to-tail sequence of individual arcs. Hence it considers the path from the latch through the bidirectional cell back to the latch. One must use path segmentation by specifying set_input_delay and set_output_delay commands. A set_output_delay or a set_input_delay command at the output of the three-state cell driving the bidirectional port breaks the timing arcs at that point. A set_output_delay command at an internal pin creates a valid endpoint and a set_input_delay creates a valid start-point.

**Scenario 6**

DC is unable to meet the timing for the path which is the worst violator. However, it does not seem to improve on other paths in the design which most certainly can be improved by merely swapping cells in those paths.

**Solution**

By default, DC creates a default path group and a clock group for each clock created. The default path group contains paths that do not terminate at a clock. Only the worst violator in each path group affects the synthesis cost function. This can be changed by using the group_path command or modifying the value of the compile_default_critical_range variable from the default of 0.0 to a larger value. This variable ensures that all paths that are within the range are considered during optimization and not only the worst path. In general, set the compile_default_critical_range variable only in the last compile step. In other words, set constraints and perform one or more compile steps until DC does not seem to improve its results. Then set this variable to a value (usually 2 or 3) then re-compile. Setting this to a large value can increase the compile time significantly.

The group_path command can be used to create explicitly a path group and specify the weight and critical_range of that group. No path can exist in more than one path group. Here is an example of a group_path command:

```
group_path -name MY_GROUP -weight 20 -critical_range 10 -to {OUT_PORT2}
```

This command creates a group MY_GROUP of all paths to the output port OUT_PORT2 with a weight of 20. The critical_range option indicates that DC should work on all paths which are within the range of 10 from the worst violator in that group.

Note that adding new path groups and defining critical ranges for path groups could result in relatively longer compilation time. Each path can exist in only one path group.

**Scenario 7**

How can one time through latches in designs as if these latches were transparent?

**Solution**

Timing through a latch is possible provided a timing arc exists between the latch data input and the latch state output. The basic approach is to disable the setup/hold timing relationship between the data input and the enable input of a latch. In order to time through all latches of a certain reference type in the technology library, the timing arc disabling can be done on the library itself. For example, the following command would allow for timing through any latch in the design whose reference is <library_name>/<latch_cell_type>.

```
set_disable_timing <library_name>/<latch_cell_type>-from  <enable_pin> -to
<data_pin>;
```

If you wish to time through a subset of latches in the design, the timing arc disabling can be performed only on that subset. Note that timing arcs for latches of different types are disabled separately. For example, if you want to time through latches latch_A, and latch_B, which have a reference name LD1, the following command would allow timing through latches A and B:

```
set_disable_timing {latch_A latch_B} -from G -to D;
```

**Scenario 8**

A top level module has a few submodules and a hand-crafted clock circuitry at the top level. You wish to synthesize this design to gates but leave the clock logic at the top level intact. How does one go about accomplishing this?

**Solution**

The dont_touch_network command will propagate the dont_touch attribute throughout the hierarchy for the clock network. Since this command specifically works for clock networks, it is required that a clock object be defined (using create_clock) before this command is used. If the intention is to only maintain the clock logic at the top level the following script can be used to set the dont_touch attribute on all leaf cells at the top level of the design which constitute the clock circuitry.

```
/*Defining variable var1 which contains list of all instances at top level */
var1 = find(cell, "*")
/* Finding instances of all hierarchical cells at top level */
filter var1 "@is_hierarchical == true"
var2 = dc_shell_status
var3 = var1 - var2
dont_touch var3
```

**Scenario 9**

How does one find all the cells of a particular reference in a hierarchical design? In other words, you have a hierarchical design with the FD1 (flip-flop) library cell used several times and you wish to get an actual count to identify if it is worth requesting a special low drive cell of the same functionality.

**Solution**

The simplest way would be to ungroup the design from the top level and use the report_reference command. Alternatively, if one prefers not to ungroup the design, a script which finds all the cell instances which reference the FD1 should accomplish the same. Then the total number of cells in this list is counted.

```
cell_list = {}
filter find(cell, "*", -hier) "@ref_name == FD1"
cell_list = dc_shell_status
count = 0
foreach (cell_instance, cell_list){
      count = count + 1
}
echo "Total number of FD1 cells in design are" count
```

**Scenario 10**

You are using assign statements in your Verilog code to assign duplicate logic to 3 different output ports. On compiling the design, instead of creating 3 sets of this duplicate logic, DC optimizes away two sets. It creates one copy of the logic and assigns it to the output ports by shorting them together. How can one create 3 copies of this logic, each going to an output port?

**Solution**

The only way to do this is to group the 3 sets of logic into 3 different levels of hierarchy before compiling the design. After compile, you can ungroup the 3 levels and you should see 3 copies of the logic, each going to an output port.

**Scenario 11**

What is the difference between '-' and 'X' in the std_logic type? Synthesis results seem to be the same regardless of which one is used.

**Solution**

The std_logic_1164 package (in $SYNOPSYS/packages/IEEE/src) contains the following:

```
 PACKAGE std_logic_1164 IS
--------------------------------------------------
   -- logic state system  (unresolved)
--------------------------------------------------
```

```
TYPE std_ulogic IS ( 'U',  -- Uninitialized
                     'X',  -- Forcing  Unknown
                     '0',  -- Forcing  0
                     '1',  -- Forcing  1
                     'Z',  -- High Impedance
                     'W',  -- Weak     Unknown
                     'L',  -- Weak     0
                     'H',  -- Weak     1
                     '-'   -- Don't care
                   );
attribute ENUM_ENCODING of std_ulogic : type is "U D 0 1 Z D 0 1 D";
```

From the ENUM_ENCODING attribute, you can see that '-' and 'X' are mapped to 'D', which is a don't-care (can be either '0' or '1'). Therefore, for synthesis, '-' is equivalent to 'X' which is equivalent to 'D'. This is the reason your synthesis results are the same, regardless of whether you use '-' or 'X'.

**Scenario 12**

You have a flop with both asynchronous reset and preset. If both of these inputs are active, how would the DC handle this design assuming that the library does not have any information on the output when both are active. In other words, you wish to know whether will DC ensure that such condition does not occur or will DC ignore it.

**Solution**

When coding for synthesis, one has to specify a priority to one of these signals that is either reset or preset. Therefore, a scenario when both these inputs are active will not occur. In v3.1a of DC, there are two new attributes, one-hot and one-cold, which can be specified in the source HDL to tell DC that these signals are mutually exclusive and DC will not synthesize priority logic. This is an unstable state, so DC will ignore it (DC does static timing analysis). There are variables you can specify in your library (clear_preset_var_1 and clear_preset_var_2) source (.lib file) which allows you to model this state, that is the output value when this state occurs. DC will still ignore it. This modeling capability is provided for simulation purposes only. If the variables are not set in the library, you will get an X at the output during simulation, when both reset and preset are active at the same time.

**Scenario 13**

What features require the use of DC-Expert instead of DC-Professional?

**Solution**

Commands requiring DC-Expert licenses are as follows:

1. Multiple Frequency Clocking

   Given: clk_a is 40 MHz and clk_b is 80 MHz

   Commands:

   ```
   create_clock -period 25 -waveform {0, 12.5} clk_a
   create_clock -period 12.5 -waveform {0, 6.25} clk_b
   ```

2. Pipeline Retiming

   Any of these commands:

   ```
   set_balance_registers [-design design_list]
   set_balance_registers true [-design design_list]
   balance_registers
   ```

3. Latch-based Time Borrowing

   Command:

   ```
   set_max_time_borrow time_limit object_list
   where time_limit > 0.0
   ```

4. Critical Path Re-synthesis

   Commands:

   ```
   group_path -critical_range critical_range_value -name path_group
   compile -map_effort high -incremental
   ```

5. In-place Optimization after layout

   Command:

   ```
   compile -in_place
   ```

6. Incremental Design Editing (that is, netlist editing)

   Any of these commands:

   ```
   create_design
   create_cell/remove_cell
   create_net/remove_net
   create_port/remove_port
   create_bus/remove_bus
   connect_net/disconnect_net/all_connected
   ```

**Scenario 14**

A design has a delay cell (instance name u1) instantiated in HDL code as shown in Figure 4.8. This delay cell is between a flip-flop and mux, and you wish to make sure DC does not remove it from this location during synthesis. The flip-flop and mux are generics. The delay cell does have a dont_touch attribute on it in the library. You tried setting a dont_touch on the output of the cell and compiled, but DC leaves the output of the cell hanging (unconnected) after compile.

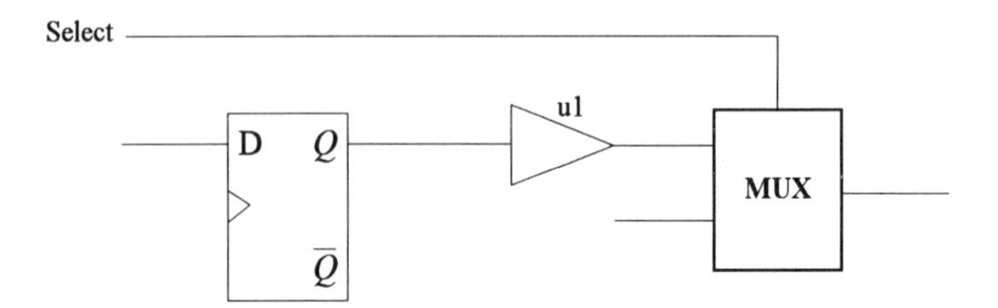


**Figure 4.8 Design Illustrating the Usage of dont_touch**

**Solution**

The dont_touch attribute is ignored on nets which have only unmapped cells on them. If the net has some mapped and some unmapped logic, the dont_touch attribute may be ignored if a better structure is found that does not require that net. Create another module containing the delay cell and instantiate this module in the hierarchy. Then place a dont_touch on the reference (that is, the delay cell module) before compiling.

**Scenario 15**

A design has an address bus 32 bits wide of which only 2 bits go into a module. You create an extra level of hierarchy in DC using the group command and only 2 bits of the address needed go into the newly created module. DC brings in all 32 bits into the module and does not connect the top 30. Is there a way to get rid of the unused bus ports?

**Solution**

One can remove unused ports using the remove_port command as shown below:

```
remove_port "a(26)"
```

The port name should be enclosed in quotes. However, there is a potential problem with this approach. Once you have removed these unused ports, the top level will not be able to link to the lower level. The reference in the top level will still include the unused ports, but the tool will not be able to find these ports in the lower level since you've removed them! One way to get around this is to execute the group command in the design that contains the unused ports.

```
group find(cell, "*") -design_name new
```

This will create a new design called new which will not contain the unused ports. You can then go into the top level design and execute ungroup on the design you just grouped. For example, if the design containing the unused ports is called A (instance name A1) and the top level is called top, you should execute the following commands:

```
current_design A
group find(cell, "*") -design_name new -cell_name new1
current_design top
ungroup A1
```

The instance new1 in top will now only contain the ports that are used.

**Scenario 16**

In a state machine process, if a state is supposed to remain the same under a certain condition, does the user have to explicitly write

```
next_state <= current_state;
```

Since nothing new is assigned to it shouldn't it maintain the state even if not specified?

**Solution**

If you do not have the next_state <= current_state statement in the combinational process statement, DC will infer latches for the next_state signal. In general, when using a case statement, one must cover all the possibilities or, in other words, assign the next_state under all conditions.

**Scenario 17**

Is there a way to control instance names inferred by DC during synthesis?

**Solution**

No, there is no way to control instance names except by adding prefixes and suffixes. This can be achieved using the following variables:

```
compile_instance_name_prefix = "U"
compile_instance_name_suffix = "S"
```

**Scenario 18**

Some of the DC warning messages are too long, elaborate and downright annoying. Is there a way to suppress information and warning messages in DC?

**Solution**

This can be done by setting the following variable to false. By default, this variable is set to true.

```
dc_shell> list verbose_messages
verbose_messages = "true"
1
```

Shown below is an example of a dc_shell session before and after setting the above variable to false.

```
dc_shell> set_input_delay 3 all_inputs()
Performing set_input_delay on port 'data'.
Performing set_input_delay on port 'clk'.
Performing set_input_delay on port 'reset'.
Performing set_input_delay on port 'read'.
1
dc_shell> verbose_messages = false
"false"
dc_shell> !set
set_input_delay 3 all_inputs()
```

However, sometimes, some warning messages are more annoying than others. (George Orwell's, "Animal Farm" comes to mind!). In such cases, specific messages can be turned off using the suppress_errors variable as shown below.

```
dc_shell> current_design
Current design is 'datagen_clks'.
{"datagen_clks"}
dc_shell> find(port, clock)
Warning: Can't find port 'clock' in design 'datagen_clks'. (UID-95)
{}
dc_shell> suppress_errors = {"UID-95"}
{"UID-95"}
dc_shell> !find
find(port, clock)
{}
```

**Scenario 19**

You have a mapped design obtained after compile but the following command:

```
get_attribute current_design is_unmapped
```

shows the unmapped attribute to be true. But the **report_cell** output shows all the cells in the design from the target vendor library.

**Solution**

This could be due to the absence of power and ground cells in the technology library. This can be verified using the following command:

```
filter find(cell, "*", -hier)  "@is_unmapped == true"
```

Check the references of the unmapped cell instances using the **get_attribute** command as shown below:

```
dc_shell > get_attribute U234 ref_name
Performing get_attribute on cell 'U324'.
{"**logic_1**"}
```

This can be overcomed by updating the technology library with power and ground cells. Create a library with the power and ground cell. Then update the technology library with the newly created library using the **update_lib** command.

## Example 4.4    Library with only the Ground Cell

```
library ("my_lib.db"){
cell (GND) {
	area : 1;
	pin (Z) {
		direction : output;
		function : "0";
	}
}
}

/* dc_shell script */
read_lib my_lib.lib
write_lib my_lib
update_lib <technology_library> my_lib
```

Example 4.4 shows a library (say my_lib.lib) with only a ground cell. Since the cell has a function attribute a Library Compiler license is required.

**Scenario 20**

You are using the embedded script feature of the VHDL-Compiler to automatically write out a pre-optimized version of VHDL. In other words, you wish to read in the source VHDL and write out a db file prior to compile via a dc_shell script embedded in the source VHDL. But you find that DC issues an error. The embedded script is as follows:

```
-- pragma dc_script_begin
-- write -output sync_mon_READ.db
-- pragma dc_script_end
```

```
Error: Can't execute command 'write' in this context on or near line 2. (EQN-14)
Error: Previous error has stopped execution of HDL script for 'sync_mon'. (UI-41
)
```

**Solution**

This is because in embedded dc_shell script, only commands that set constraints and attributes are allowed. Commands such as compile, ungroup, or report cannot be used in embedded scripts.

**Scenario 21**

You wish to replace an instance (U1) of the INV library cell in the netlist by the INV2 library cell. How does one go about doing this in DC?

**Solution**

The change_link command can be used to do this. This command is a very useful command that allows the user to change the reference of a cell provided the interface (the number of inputs and outputs) of the two cells match. If the interfaces of the two cells do not match, create a Verilog (VHDL) module (entity) with a matching interface for change_link.

```
change_link find(cell,"U1") <target_library>/INV2
```

## Recommended Further Readings

1. Design Compiler Family Reference Manual
2. DesignTime: Constraining Designs for Timing and Analysis Application Note
3. Flattening and Structuring: A Look at Optimization Strategies Application Note
4. Synopsys Newsletter, *Impact* Support Center Q&A

CHAPTER 5

# Constraining and Optimizing Designs-II

This chapter is a continuation of logic synthesis strategies using DC. First, FSM synthesis steps are outlined using examples from Chapter 2. Then, the tips for FSM synthesis are provided. This is followed by a discussion on fixing min_delay violations during synthesis. One of the biggest advantages of logic synthesis is the ability to target different technology libraries. Technology translation is discussed and the steps involved in translating designs with black-boxes are outlined. Finally, a number of classic scenarios have been discussed.

## 5.1 Finite State Machine (FSM) Synthesis

Finite State Machine synthesis involves a number of steps. This section provides a description of the steps in DC after coding the source VHDL.

### Example 5.1 Code for Finite State Machine

```
entity test is
port (X, clock : in bit;
              Z : out bit);
end test;
architecture trial of test is
signal ST : state;
begin
process
    begin
     wait until clock' event and clock = '1';
     if  X='0' then Z = 0;
```

```
else
   case ST is
      when S0 =>
            ST <= S1 ;
             Z <= '0';
      when S1 =>
            ST <= S2 ;
             Z <= '0';
       when S2 =>
            ST <= S3;
              Z <= '0';
        when S3 =>
            ST <= S0 ;
             Z <= '1';
   end case;
end if;
end process;
end trial;
```

The steps involved between writing the source code and generating the required state table representation of the state machine are as follows. First the HDL source code is mapped to cells from a target technology library. Then, the flip-flops in the design which hold the current state of the FSM must be identified. The next step involves assigning specific codes to different states. This is followed by a grouping of the state flip-flops and their associated combinational logic into a separate level of hierarchy. Grouping helps isolate the FSM from the rest of the design. Once grouped into a separate level of hierarchy, this sub-design can now be represented as a state table.

## SCRIPT TO READ, COMPILE AND EXTRACT FSM

```
read -f vhdl mealy.vhd
create_clock -period 10 clock -waveform {0 5}
compile -map_effort low
set_fsm_state_vector {ST_reg[0] ST_reg[1]}
set_fsm_encoding { "S0=2#00" "S1=2#10" "S2=2#01" "S3=2#11" }
group -fsm -design_name eg1_fsm
current_design = eg1_fsm
report_fsm
extract
report_fsm
```

```
/*Output of report_fsm*/
******************************************
Report : fsm
Design : eg1_fsm
******************************************

Clock         : clock     Sense: rising_edge
Asynchronous Reset: Unspecified

Encoding Bit Length: 2
Encoding style     : Unspecified

State Vector: { ST_reg[0] ST_reg[1] }

State Encodings and Order:
S0        : 00
S1        : 10
S2        : 01
S3        : 11
```

The script reads the VHDL code into DC, compiles the design and extracts the state machine. The **compile** command maps the HDL code to target technology library cells. At this stage, DC is unaware that the VHDL description is a state machine. To verify if DC understands that the design is a state machine, the **report_fsm** command can be used. Identifying the state vectors using the **set_fsm_state_vector** command, tells DC that the design is a state machine. Further, setting the state vectors helps the tool differentiate the state flip-flops from the flip-flops used for registered outputs. DC by default, assigns the name, signal name (ST in this example) followed by _reg[i] (ST_reg[i] for this example), where i is the number of state vector bits, to the state vectors. In this case, i takes the values 0 and 1 since only two bits are essential.

The **set_fsm_encoding** command allows the designer control over the state encoding. While several encoding styles for FSMs exist, we will discuss the auto (default encoding style) and one-hot encoding styles. The encoding style can be assigned using the **set_fsm_encoding_style** command. The **group -fsm** command groups the state flip-flops and the associated combinational logic into a separate level of hierarchy. On extraction, the state machine can be written out in state table format.

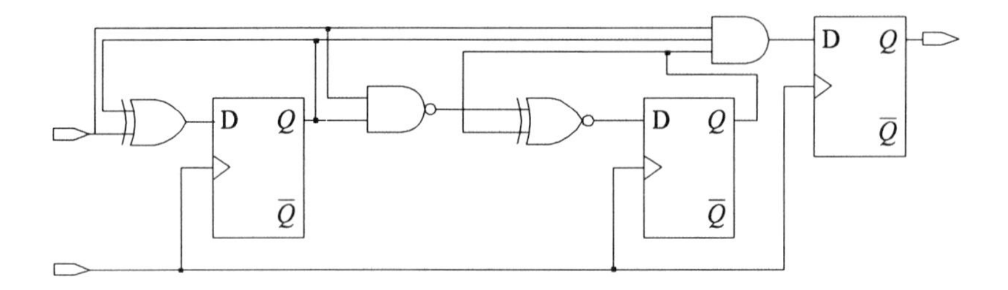


**Figure 5.1 Top Level Design After Synthesis**

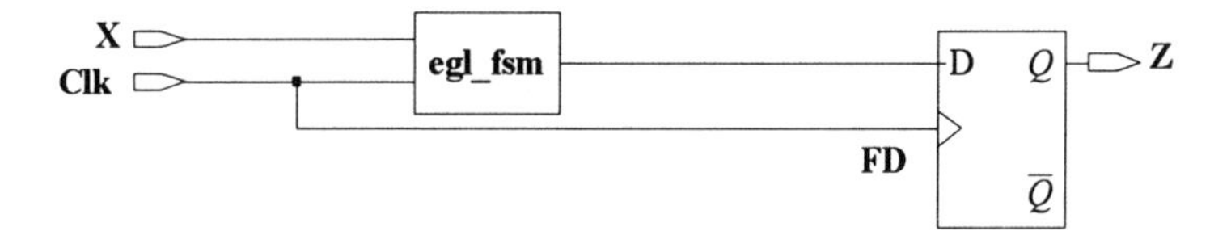


**Figure 5.2 Top Level Design After Grouping**

Figure 5.1shows the top level design generated from the VHDL code. Notice that there are three flip-flops. Figure 5.2 shows the design after grouping and extraction. One of the flip-flops was not a part of the state machine and hence was not grouped under the FSM block eg1_fsm. This flip flop is an output flip flop implying a registered output.

## Example 5.2 VHDL Code for FSM and Synthesis script

```
package fsm_states is
type state is (S0, S1, S2, S3); -- state can take one of these values.
  attribute ENUM_ENCODING : STRING;
  attribute ENUM_ENCODING of state : type is "00 01 10 11";
end fsm_states;


library work;
use work.fsm_states.all ;
entity test is
port (X, clock : in bit;
```

```
    Z : out bit);
end test;
architecture trial of test is
  attribute STATE_VECTOR : string;
  signal ST : state;
  attribute STATE_VECTOR of trial : architecture is "ST";
begin
process
    begin
       wait until clock' event and clock = '1';
        if  X='0'  then  Z <= '0';
        else
           case ST is
                when S0 =>
                          ST <= S1 ;
                           Z <= '0';
                 when S1 =>
                          ST <= S2 ;
                           Z <= '0';
                  when S2 =>
                           ST <= S3;
                           Z <= '0';
                  when S3 =>
                            ST <= S0 ;
                             Z <= '1';
            end case;
     end if;
end process;
end trial;
```

## SCRIPT TO READ, COMPILE AND EXTRACT FSM

```
read -f vhdl one-hot.vhd
report_fsm
compile
group -fsm -design_name fsm_1hot
current_design = fsm_1hot
extract
set_fsm_encoding_style one_hot
set_fsm_encoding {"S0=2#1000" "S1=2#0100" "S2=2#0010" "S3=2#0001"}
```

```
set_fsm_minimize  true
compile
current_design = fsm_1hot
report_fsm
extract
write -f st fsm_1hot -o one-hot.st

*******************************************
Report : fsm
Design : fsm_1hot
*******************************************

Clock             : clock      Sense: rising_edge
Asynchronous Reset: Unspecified
Encoding Bit Length: 4
Encoding style     : one_hot
State Vector: { ST_reg[3] ST_reg[2] ST_reg[1] ST_reg[0] }
State Encodings and Order:

S0        : 1000
S1        : 0100
S2        : 0010
S3        : 0001
```

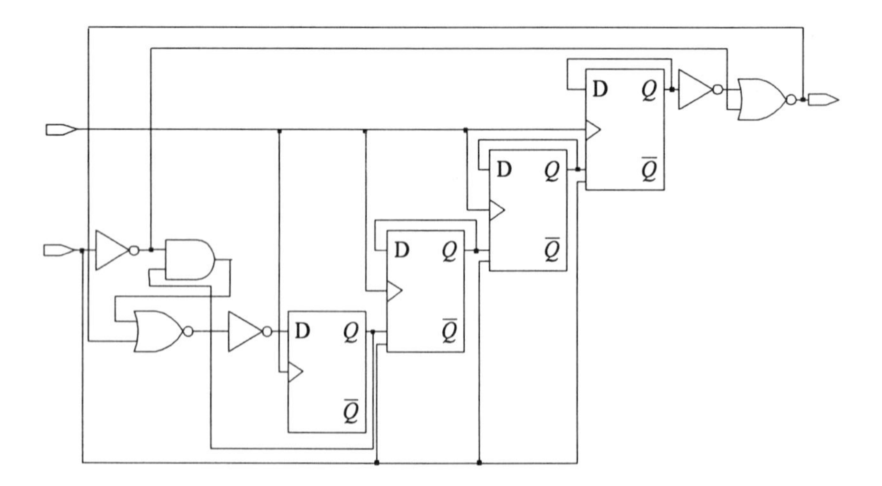

**Figure 5.3 One-Hot FSM**

In the dc_shell script file used to synthesize the FSM, notice that the script changes the encoding style to one-hot inspite of the encoding being different in the source code. Figure 5.3 shows the FSM inferred. The same dc_shell script can be used for Verilog code along with the appropriate file names.

## Tips for FSM synthesis

1. Before a state machine has been extracted and after the group command, DC sometimes fails to group some of the surrounding logic which would then have made the state machine logically more optimal. The cells which are grouped are those in the transitive fan in/out of the state vector cells. The group command does not, however, include cells which do not completely fanout/in to the FSM. In other words, after grouping, one might find two inputs to this grouped level of hierarchy which are the opposite (inverted) of each other. In this case, one must use the characterize -connections command with the current_design set to the top level, so that the connection attributes are passed on to the newly grouped level of hierarchy. Another method is to use the set_opposite command which tells DC that the two pins are in fact opposite to each other. The dc_shell script to characterize is as follows:

   ```
   current_design = TOP
   characterize -connections fsm_block
   current_design = fsm_block
   extract
   ```

2. Once the state machine is extracted, the design can be written out in state machine format or to the original RTL VHDL format by the following steps:

   ```
   current_design = fsm_block
   write -f vhdl fsm_block -o rtl.vhd
   ```

3. One constant complaint among users is the lack of control over net names, especially when the concerned nets are inputs to the state machine. One possible work around for this is to set the variable, write_name_nets_same_as_ports to true (this is false by default) before writing the design in EDIF. Then read the EDIF back into DC and follow the steps till extraction of the FSM. After reading in the edif file, the ports and the nets connected to them should have same names. It is advisable to do a compare_design between the new design read in and the old design in memory, to ensure that no changes occurred during the write step.

4. reduce_fsm and set_fsm_minimize are two commands users tend to confuse. reduce_fsm is a command while set_fsm_minimize is a switch. reduce_fsm should be executed after the extract command to reduce the transition logic between states. The set_fsm_minimize is turned on prior to compile so that the tool infers the minimum number of states required for the fsm. This command can be used along with the set_fsm_preserve_state command to maintain certain states during compile.

5. Last but not the least, efforts must be made to clearly partition the design into control logic and data path elements. Reading in a large netlist and executing the extract command is not an effective methodology.

## 5.2 Fixing Min Delay Violations

Once the max_delay requirements imposed due to the setup constraints for the sequential cells have been met, DC can be used to fix the minimum path delay requirements. Since the path delays are the maximum in the *worst case* timing analysis or *worst case* operating conditions, max delay requirements must be met in the *worst case* operating conditions.

The minimum delay requirements are set by the hold constraints for the sequential cells. Hold time problems are caused due to short delay paths between registers which cause the data signal to propagate through two adjacent flip-flops on a single clock edge. Since path delays are the shortest under *best-case* operating conditions, hold time problems are maximum in these conditions. Hence, hold violations have to be fixed under these conditions.

One approach to go about fixing both the setup and hold constraints is a two-pass compile approach. In the first pass compile, fix the setup violations under the *worst case* operating conditions. Then set the operating conditions to *best-case* for the second pass compile. Use the fix_hold command to set an attribute fix_hold on the clock objects for which hold constraints have to be met. The second pass compile should be with the compile switch -only_design_rules turned on. This should fix all the hold violations in your design. Also since under the "best-case" operating conditions, the max_delay paths will have excessive positive slack, hold constraints maybe fixed at the cost of setup constraints. Such a situation can be avoided by adjusting the constraints such that the critical paths in the design appear critical under best-case conditions. One of the ways to achieve that is by specifying a negative uncertainty on the clock by using the set_clock_skew -minus_uncertainty command.

Hold time problems will generally occur in shift register structures or scan chains. Since by default DC treats the clock as ideal with no path delays, one must account for the network delay by using the set_clock_skew -propagated command.

## 5.3 Technology Translation

Conversion of a design netlist from one technology library to another is called technology translation. This powerful capability helps compare performance across different ASIC vendor libraries. However, this utility has its limitations and works best with combinational logic. Technology libraries differ in the cells they contain and in area and timing. Hence, after technology translation, optimization must be performed on the design to meet the original design constraints.

In DC, technology translation is performed by the translate command. In order to perform translation from one technology to another, the first requirement is the availability of both the existing library to which the netlist has been mapped and the target_library. Shown below are the steps involved in translating a design top from technology library libA to technology library libB.

```
current_design = top
target_library = libB
link_library = libA
search_path = search_path + "path to the two libraries"
translate
```

The translate command replaces each cell in the design with the closest matching functional cell from the target_library. In case such a matching cell is not found then it is converted to a cell from the generic library. The dont_use, dont_touch, set_default_register_type and the prefer command are useful commands that affect the translation process.

## 5.4 Translating Designs with Black-Box Cells

When translating designs from one library to another, DC performs an instance by instance functional comparison. In other words, translation requires that a cell in the current technology library and the target technology library both have the same functionality, if an instance of the cell is to be translated. The translate command does not translate black-box cells during technology translation. However, there exists a simple trick to translate a black box cell. *Black-box cells are cells with a function attribute which cannot currently be described in the Synopsys Library Compiler syntax or those which do not have a function attribute specified.* Such cells have the b attribute attached to them implying a black box cell. The report_lib command can be used to identify all the attributes on the cells in the library as shown below:

```
report_lib libA
```

It is possible that the design prior to translation has one such cell instantiated in it and the target library does contain an identical cell. Since DC does not see any functionality described, it is unable to translate this particular instance. A design with black-box cells can be translated by the following steps:

1. Identify the black-box cells in your design and then find the equivalent cells in the target_library.
2. Create a translation library for these black-box cells. For example, if your netlist has a black-box cell mem and your target_library contains an equivalent cell mem_new, then create a translation library which is essentially a module that instantiates the target cell mem_new, but with the same interface as the mem cell as shown in Example 5.3.

### Example 5.3 VHDL/Verilog Translation Library

```
module mem (D0, D1, D2, D3, clk, Q0, Q1, Q2, Q3);
input D0, D1, D2, D3, clk;
output Q0, Q1, Q2, Q3;
mem_new I1 (.I0(D0), .I1(D1), .I2(D2), .I3(D3), .G(clk), .Q0(Q0), .Q1(Q1), .Q1(Q1),
.Q2(Q2), .Q3(Q3));
endmodule
```

**VHDL Translation Library**

```
entity mem is
  port (D0, D1, D2, D3, clk: in std_logic;
        Q0, Q1, Q2, Q3: out std_logic);
end mem;
architecture STRUCTURAL_VIEW of mem is
  component mem_new
  port (I0, I1, I2, I3, G: in std_logic;
        Q0, Q1, Q2, Q3: out std_logic);
end component;
begin -- using positional association
I1: mem_new port map (D0, D1, D2, D3, clk, Q0, Q1, Q2, Q3);
end STRUCTURAL_VIEW;
```

3. After the translation library has been created, convert the design to the db format using the read and the write commands. You have now created a block around the cell in the target_library with an interface similar to the interface of the black-box

cell in the current library netlist. Assuming that your translation library is called translation.db, your original library original.db, and your new target technology library new.db, set the link_library variable as follows:

```
link_library ={translation.db original.db new.db}
```

Also, ensure that the search_path variable points to all the directories containing these libraries.

4. Execute the link command to translate the black-box cells in the netlist to the library new.db. During the link operation, DC checks the link_library for cells beginning with the translation.db library, followed by original.db and finally, the new.db. On finding the mem design in translation.db, it links to the newly created design. During translation, the mem cell is nothing but a sub-block with the mem_new instantiated in it, and mem_new is a cell in the target_library new.db. This level of hierarchy can later be removed with the ungroup command.

If there is no exact equivalent cell in the target library, you can create a structural model of the black box cell using primitives from the target technology library. Then, as in the above case, create a translation library with the same interface as the black box.

# 5.5 Pad Synthesis

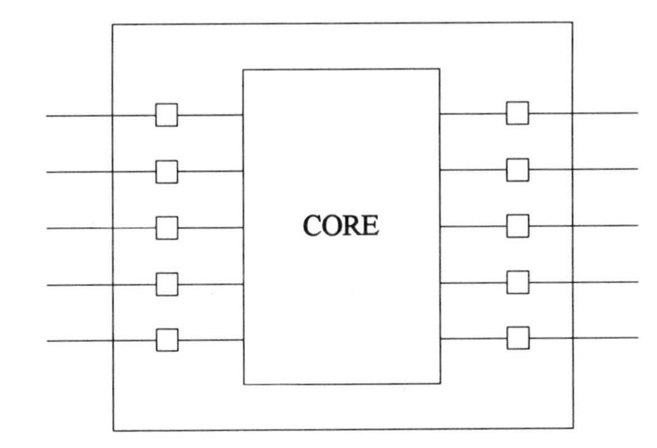


**Figure 5.4 ASIC Showing Core and Pad Cells**

Adding pads to your design is an essential part of the design process. One option is to instantiate pads after the core of the design has been implemented and simulated. Figure 5.4 shows an ASIC core with the pad cells. DC provides a means for automatic pad insertion. However, this is entirely dependent on the ASIC vendor library having appropriately modeled pad cells. A pad cell in the Synopsys library is one which has the pad_cell attribute set to true. Also, one or more of the pins of the pad cell will have the is_pad attribute set on them. Hence, the first step to attempting pad synthesis is to ensure that the technology library has pad cells modeled appropriately. The following commands can be used to determine all the pad cells in the technology library:

```
filter find (cell,libA.db/*) "@pad_cell == true"
```

If DC issues a message that it is unable to find the library libA.db, execute the list -libraries command at the dc_shell prompt. This command should list the UNIX file name and the actual name of the library. Having determined the pad cells available in the library, the next step is to find the pin on the pad cell (say padA) that has the is_pad attribute. This can be done using the following command:

```
dc_shell> filter find(pin,"libA.db/padA/*") "@is_pad == true"
Performing filter on port 'A'.
Performing filter on port 'GZ'.
Performing filter on port 'Y'.
{"Y"}
```

Having determined the pad cells available in the technology library, the next step involves pad insertion. This is done using the insert_pads command. However, if you wish to control the kind of pad cell inserted by DC, this can be achieved using the set_pad_type command. This command controls the attributes and properties of the pad cell synthesized by DC. To provide greater control, the set_pad_type command has a -exact option which helps the user explicitly specify the pad cell to be inserted from the library. The insert_pads command does not bus together inputs into the same pad using bused pad cells. Such pad cells will have to be instantiated. DC does not map to pad cells during the regular compile if the pad cells have required attributes.

## 5.6 Classic Scenarios

### Scenario 1

You are performing trial compile runs. You do not wish that wire loads be considered in these trial runs. Can one prevent DC from selecting a wire_load model for a design, or does it default to a particular wire load model?

**Solution**

The different mechanisms of selecting the wire_load model in DC is described in chapter 6. However, to prevent the use of any wire-load model, one must perform the following steps. Set the variable auto_wire_load_selection to false. Also, if the ASIC vendor library has the attribute default_wire_load set to a particular wire_load model, the following command must be used to remove the default_wire_load attribute:

```
remove_attribute library_name default_wire_load
```

**Scenario 2**

Your design has a number of internally generated signals which drive the enable pins of latches. For example, you have a state machine generated signal which drives the enable pin of a latch. The output of the latch drives a block of combinational logic, which in turn drives a primary output as shown in Figure 5.5. The time delay in the signal reaching the primary output is dependent on how soon the enable signal can be generated, and the delay through the combinational logic, after the data is latched. You wish to constrain the entire path along the enable line to the primary output.

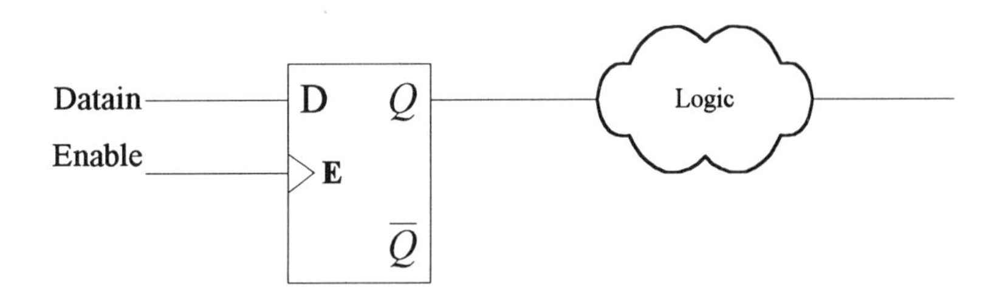


**Figure 5.5 Latch Driving Primary Output Through Combinational Logic**

**Solution**

There is no constraints on the enable (clock) line since there are no setup requirements on the clock pin. Consider a two step constraint approach using the set_output_delay and the set_max_delay commands. The path from the enable pin of the latch to the primary output can be constrained using the set_output_delay command. The path to the enable pin of the latch can be constrained using the set_max_delay command.

**Scenario 3**

After performing an initial place and route, you have available accurate load values on the nets in the design. You specify these loads on the appropriate nets in the design using a set_load dc_shell script. You then perform in-place optimization, but find that

DC does not seem to consider the new load specified on that net and appropriately size the driver cell. For example, in Figure 5.6, you expect DC to use a higher drive AND gate after in-place optimization.

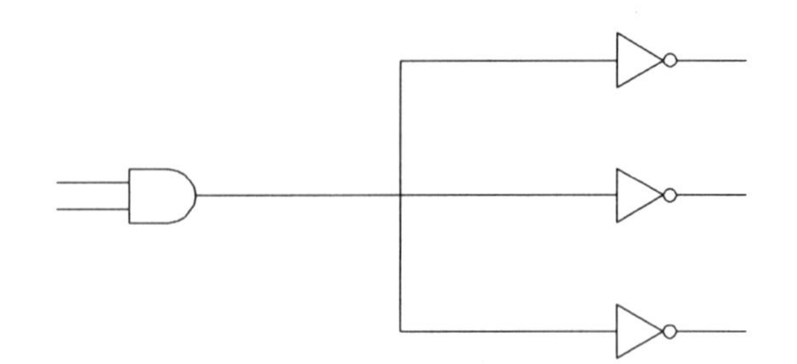

**Figure 5.6 AND Cell Driving Three Inverter Loads**

**Solution**

It is likely that no max_transition or max_capacitance constraints were set on the design. These constraints can be default values set in the library or explicitly set by the designer. Using only the max_fanout constraint could be a possible reason for DC not upsizing the instance of the AND cell. The max_fanout constraint considers only the fanout loads on the pins driven by the cell and does not consider the loads on the net (Refer to section 4.1.1). Specify max_transition or max_capacitance constraints in addition to max_fanout constraints.

**Scenario 4**

DC does not infer logic that you would like it to. So you instantiate certain cells in your design. But after compile you find that DC has replaced the instantiated cells with other logic. The logic inferred by DC is correct. But you wish that DC not optimize away the cells during compile. How does one go about doing this?

**Solution**

This can be accomplished by assigning the dont_touch attribute to the instantiated cells in the design prior to compile. You can assign a dont_touch attribute to all the instantiated cells in your design using the following two commands:

```
filter find(cell, "*", -hier) "@is_mapped == true"
dont_touch dc_shell_status
```

**Scenario 5**

You have a hierarchical design as shown in Figure 5.7. The current_design is set to TOP, the top level of your design. You execute the report_constraints -all_violators command and find a number of max_fanout violations. You believe that characterizing a particular sub-block A and re-compiling that block, should fix a large number of max_fanout violations. However, after characterizing the subdesign A and compiling A, you find that none of the fanout violations seen at the top level were fixed.

```
current_design = TOP
characterize U1 /* where u1 is an instance of the sub-design, A */
current_design = A
compile
```

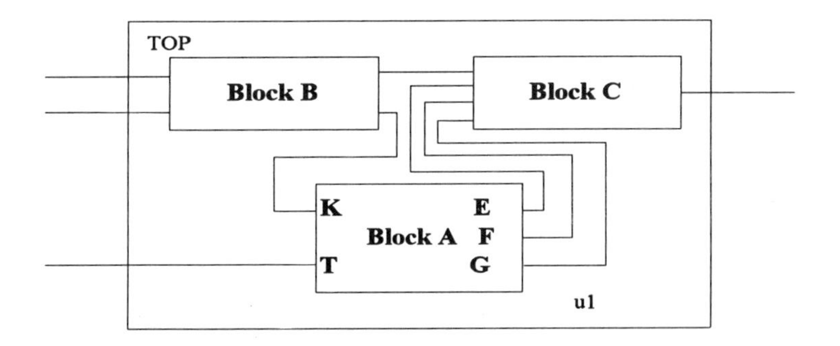


**Figure 5.7 Block Diagram Showing a Hierarchical Design**

**Solution**

The fanout violations seen at the top level were not fixed on compiling the characterized sub-block because no fanout_load values were applied to the output ports of the lower level. In other words, characterize does not capture the fanout_load drive capability required by the output ports E, F and G in Figure 5.7. The values that were applied by characterize, were load values which are not taken into account when fixing max_fanout violations (Refer to section 4.1.1 for max_fanout). Characterize command will capture this information if one were to use the characterize -constraints command instead of just characterize. This will ensure that the fanout_load values are passed down in addition to the load values on the nets.

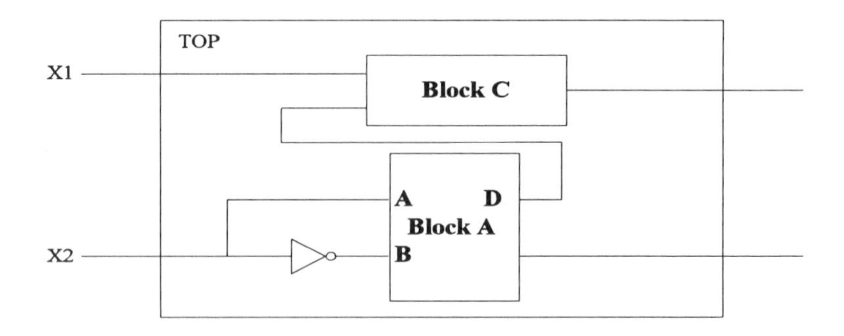


**Figure 5.8 Design with Opposite Inputs Driving a Sub-Block**

The characterize command also has another useful option, namely, connections. This is useful in a scenario where two inputs are identical except that one of them is an inversion of the other as shown in Figure 5.8. Input X2 is inverted and drives Block A at pins A and B. This information is captured when Block A is characterized with the -connections option. One can explicitly specify that two ports are opposite of each other using the set_opposite command.

**Scenario 6**

You have read in your design. DC shows inferred DesignWare parts named, "DW*". How does one ungroup the DesignWare part inferred? Even after compiling the design, the DesignWare parts exist as separate levels of hierarchy.

**Solution**

DesignWare parts inferred prior to compile are not ungrouped during compile if they are greater than four bits in size. If you have merely read in your source VHDL and DC has inferred certain DesignWare parts (you can confirm this by using the report_reference command after reading in the design), ungrouping the DesignWare parts can be accomplished by the following command:

```
replace_synthetic -ungroup
```

If replace_synthetic -ungroup is executed, high-level optimization features such as resource sharing and implementation selection will not occur since the DesignWare parts have been ungrouped. If the design is already mapped, one must ungroup the DesignWare parts based on their instance names.

**Scenario 7**

You wish to find all data pins of latches in your design. Is there a single command which will accomplish this?

**Solution**

This can be accomplished by the following command:

```
all_registers (-level_sensitive_devices -data_pins)
```

**Scenario 8**

You have two clocks in your design as shown in Figure 5.9? One is the chip input clock clk1 with a period of 10 ns (waveform {0, 5}). The other clock, clk2 is internally generated from clk1 and is the output of a flop. To account for the clk to Q delay, clk2 is specified with a period of 20 ns and a waveform of {3, 13}. In the design, the clock, clk2, goes to the clock pins of other flops. When timing the design from clock domain clk1 to clock domain clk2, DC performs a setup check from the edge at time 0 of clk1 to the edge at time 3 of clock clk2. What is the way to prevent that and only have the DC check for hold from time 0 to time 3 in the two clock domains?

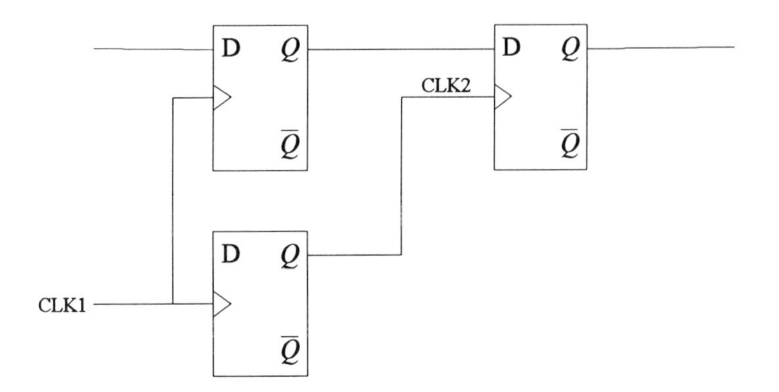


**Figure 5.9 Internally Generated Clocks**

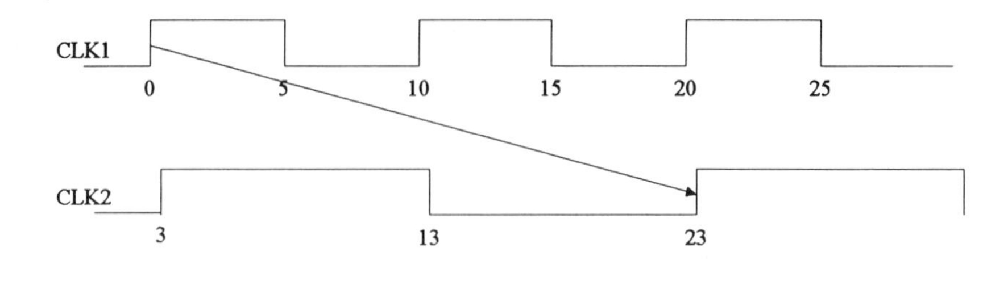


**Figure 5.10 Clock Waveforms**

**Solution**

To specify clock delay or skew on the clock line, use the command set_clock_skew. The options available with this command are -delay which specifies an absolute delay, -propagated which causes DC to dynamically calculate the delay on the clock line, and the -uncertainty option which specifies the plus or minus skew possible on the clock line. By default, DC will always perform the most restrictive setup check which, in this case, is from 0 to 3. The clock waveforms must be aligned for correct setup check from 10 to 20 as shown in Figure 5.10. Another alternative, if you wish to maintain the clock waveforms is to use the following command:

```
set_multi_cycle path 2 -setup -from clk1 -to clk2.
```

**Scenario 9**

You have a sample design containing an instantiated NAND gate. The design contains only this cell and you are trying to understand timing reports in DC. The output capacitance is set to 0, and there is no wire load set. You have set the wire load model to none to avoid any wire delays. When you execute report_timing, you get a timing delay of 3.43ns. You know that the intrinsic delay of the NAND gate is 3 ns. How does one explain the additional delay of 0.43 ns?

**Solution**

The target technology library might have a default_wire_load attribute specified. What this means is that if no wire_load is set, each time the report_timing command is executed, the default_wire_load is re-applied. So DC, is in fact, setting the wire load to none, but when report_timing is executed, it is immediately re-applied. In order to prevent this from occurring, you should perform the following steps:

```
remove_attribute lib_name default_wire_load
```

```
set_wire_load   /*(no arguments are specified to set the wire load to "null")*/
```

This must be done during each new session of the dc_shell. The alternative is to modify the library (remove the default_wire_load attribute in the .lib and write out the .db file from DC) to do this permanently.

**Scenario 10**

You have several instances in your design which have dont_touch attributes placed on them. You now wish to ungroup them, but are unable to remove the dont_touch attribute on an instance using the remove_attribute command.

**Solution**

It is likely that the instance has inherited the dont_touch attribute from its reference. If this is indeed the case, you should first remove the dont_touch attribute from the reference. Use the remove_attribute command with the find command as follows:

```
remove_attribute find(design,xxx) dont_touch
remove_attribute find(reference, yyy) dont_touch
```

This helps explicitly specify whether you are referring to a design, instance, pin, or net.

**Scenario 11**

Your technology library has a default_max_fanout specified. But DC on synthesis does not seem to buffer your clock line accordingly.

**Solution**

The default_max_fanout attribute in a library does not direct DC to buffer input ports. Since DC does not have any information on the cell driving the input ports it does not buffer the line. You should set_max_fanout on either the design or the input ports that you wish to buffer, and then optimize using the following command:

```
compile -only_design_rules
```

In general, synthesis of clock trees is not recommended. While instantiation of clock trees is used by some, others work with their ASIC vendors to devise a clock tree strategy.

**Scenario 12**

You have specified all the constraints to your design. Each clock must now have a separate path group. The remaining paths must all be in the default path group. You now wish to verify if any paths which you intend to be in a clock path group, are in the default path group instead. In other words, you wish to identify any flops in the design which might be unconstrained.

**Solution**

Use the following dc_shell script to verify.

```
/* find all flops in your design */
all_flops = {}
clk1_list = {}
my_goal = {}
all_registers(-edge_triggered)
all_flops = dc_shell_status
/* Then for all your clocks in your design find the flops */
all_registers(-edge_triggered -clock clk1)
clk1_list = dc_shell_status
my_goal = all_flops - clk1_list
```

If you wish to find all the unconstrained points in the form of a report, use the check_timing command. This command, when executed, will show all the unconstrained end points and the clock gating points, if any. The above approach can also be used for latches by using -level_sensitive instead of the edge_triggered option.

**Scenario 13**

After compiling your design you find that instances intended to be flip-flops are not mapped to technology library cells after compile. Instead, they remain as DC generic SEQGEN cells.

**Solution**

A typical case of this is when you have internally generated divided clocks. To define these clocks, it is likely that you have created a clock object on a pin of a generic cell. This will result in an implicit dont_touch attribute on that cell which prevents the cell from being mapped. In order to avoid this problem, you should first map the design to gates, before creating the clock object. After the design has been mapped, the clock object can be created and the design re-compiled. Another alternative is to create a hierarchy around the SEQGEN cell, and create_clock on the hierarchical pin connected to the SEQGEN.

**Scenario 14**

A net in your design fans out to several cells in the design. You wish to find the driver of the net. What is the best way to identify the pin driving this cell?

**Solution**

Use the following dc_shell script to find the pin driving the net.

```
all_connected find(net, N2390)
filter dc_shell_status "@pin_direction == out"
```

**Scenario 15**

After inserting pads using the insert_pads command you find clock pads inserted for some of the inputs.

**Solution**

Clock pads should normally be inserted only for the ports with a clock object created on it. However, they might be inserted on other ports if those ports are part of clock gating logic. If the pads are being inserted on a compiled netlist that contains clock enable buffers, then those ports connected to the clock enable buffers may have clock pads inserted on them also. For other regular inputs, clock pads should not be used. This problem can be avoided by specifying the set_pad_type -no_clock attribute on all inputs, except the clock input, prior to pad insertion.

**Scenario 16**

Your design uses a large amount of generate statements to instantiate the sub-blocks. Each instance has 16 generic parameters and the names are incredibly long.

1. Can one change the naming style so that the blocks are created with shortened names?
2. If the above cannot be done, can one rename the designs?

Shown below is an example of a long design name generated by DC:

```
ctrl_gen1_rst_val00_rst_val10_rst_val20_rst_val30_rst_val40_rst_val50_rst_val60_rs
t_val70_rst_val80_rst_val90_rst_val100_rst_val110_rst_val120_rst_val130_rst_val14
0_rst_val150
```

**Solution**

The following three variables allow you to customize the design names:

- template_naming_style
- template_parameter_style

- template_separator_style

For more information on these variables type help variable_name at the dc_shell prompt. To avoid all separator characters, set these variables as follows:

```
template_naming_style = "%s%p"
template_separator_style = ""
```

If the number of characters in the design name is an issue, you can use rename_design, but you would also have to change the references in your top level design.

**Scenario 17**

You have inherited a design from another designer (sounds familiar?). You are attempting to analyze the report generated by the report_constraints -all_violators command. The clock period specified is 20ns. You see the following violations:

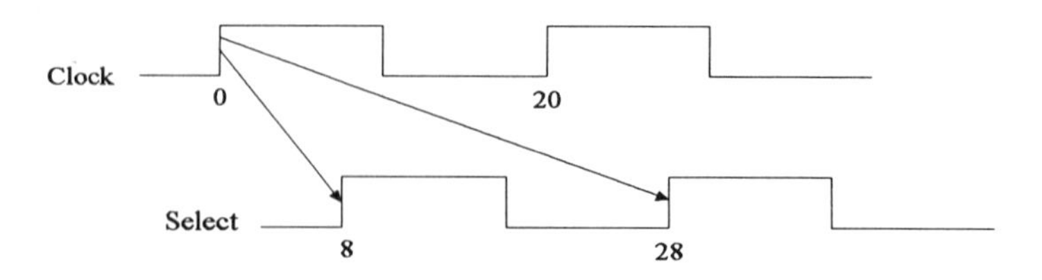


**Figure 5.11 Waveforms Showing Clock and Select Lines**

```
Startpoint: INA[5] (input port)
Endpoint: UPREAD[24] (output port clocked by SELECT)
Path Group: SELECT
Path Type: min
Point                                  Incr      Path
-----------------------------------------------------------------------
clock (input port clock) (rise edge)   0.00      0.00
clock network delay (ideal)            0.00      0.00
input external delay                   0.00      0.00 r
INA[5] (in)                            0.00      0.00 r
```

```
U324/Q (X0_NOR4)                0.08      0.08 f
U465/Q (X1_TRI1)                1.65      1.72 f
UPREAD[24] (inout)              0.00      1.72 f
data arrival time                 1.72

clock SELECT (rise edge)        8.00      8.00
clock network delay (ideal)     0.00      8.00
clock uncertainty               1.00      9.00
output external delay           0.00      9.00
data required time                9.00
------------------------------------------------------------------
data required time                          9.00
data arrival time                          -1.72
------------------------------------------------------------------
slack (VIOLATED)                           -7.28
```

**Solution**

The timing report indicates that the design has two clocks specified. One of the clocks is named clock with a rising edge at 0 and the other clock is select with a rising edge at 8. Also, the timing report shows a combinational path from INA[5] to UPREAD[24]. From the reports, it can be seen that the input port is referenced with respect to the clock and the output port with respect to clock select. The set_input_delay and the set_output_delay commands must have been specified for the input and output ports. Also, the input external and output external delay are specified as 0 indicating that a 0 value must have been specified with the set_input_delay and set_output_delay commands.

The report indicates that the hold check is being performed between edges 0 and 8 of clocks clock and select respectively. Note that when a set_output_delay command is specified, DC tries to check for both setup and hold. One of the possibilities for such a violation is that the setup check is being done between 0 and 28 edges of clocks clock and select. The hold check from 0 to 8 is equivalent to that between 20 and 28. Hold and setup checks are performed at the same edges at the endpoint, but the hold is moved one cycle from the setup launch edge of the source clock. Hence, the setup check is from 0 to 28 ns. This can be accomplished by the following command:

```
set_multicycle_path 2 -from clock -to select
```

**Scenario 18**

You execute the check_design command. DC issues connection_class violations at all the clock pins of flops in the design. What are they and what is the significance?

**Solution**

In general, DC assumes a connection_class of default for all ports. ASIC vendors sometimes provide a connection class attribute on the clock pins of flip flops. DC, during compile, ensures that if the output of a cell is connected to the clock pin of a flop which has a specific connection class attached to it, the tool requires that the output pin of the cell have the same connection class. It is possible that in this case, the clock pin does have a specific connection class and the port connected to it has a default connection class. Hence, class violations are reported. This error can also occur if you have instantiated your clock buffers with connection class mismatches and placed a dont_touch_network on the clock.

In general, the connection class attribute can be used very effectively to control clock buffering. Connection class violations are fixed during the DRC fixing phase of the compile.

**Scenario 19**

You have a hierarchical design for which you wish to place a dont_touch on all the designs at the lower level. How does you go about doing this?

**Solution**

The following dc_shell script can be used to do this.

```
current_design = A
filter find(cell, "*", -hier) "@is_hierarchical == true"
get_attribute dc_shell_status ref_name
dont_touch dc_shell_status
```

**Scenario 20**

You are using the balance_buffer command to create a balanced buffer tree for the scan enable port. But balance_buffer does not seem to add any buffer tree. Why?

**Solution**

The balance_buffer command is constraint driven. Adding the buffer tree causes an increase in the area. If the scan enable port is constrained with say, a max_transition, then DC inserts the buffer tree to ensure that this constraint is met. Also note that the balance_buffer command does not create a tree through a hierarchy. The following command can be used to specify a max_transition constraint on the scan enable port:

```
set_max_transition 1.0 scan_enable
```

**Scenario 21**

How does one specify the slew rates for the input ports of the design.

**Solution**

There is no command available to do that. DC uses the drive information of the input ports specified, using either the set_driving_cell or the set_drive command, and calculates the transition time of the ports.

**Scenario 22**

You have a gated clock in your design. You wish to specify infinite drive on an internal pin.

**Solution**

There is no command to specify infinite drive on an internal pin. Infinite drive can only be specified on primary inputs. Use the set_clock_skew -delay <value> command to specify the actual delay on the clock network. This is usually provided by the ASIC vendor. This will prevent DC from calculating the delay on the clock line, due to transition delays and loading.

**Scenario 23**

You have two clocks in your design: CLK1 (20 ns clock period) and CLK2(10 ns clock period). There are datapaths in the design between the two clock domains. The functionality of the design is such that the clock domain CLK2 captures new data only when CLK1 is low. How is the information passed on to DC for optimization?

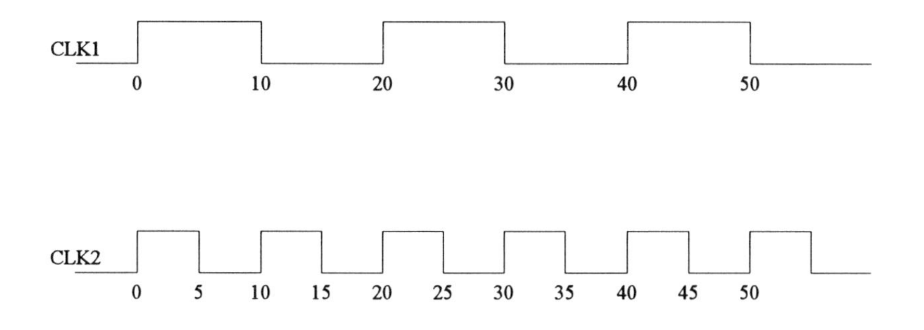


**Figure 5.12 Clock Waveforms**

**Solution**

By default, DC performs a restrictive setup check between time 0 (CLK1) and 10 (CLK2) as shown in Figure 5-13. Since new data is captured by CLK2 only when CLK1 is low, this gets translated to two cycle paths between clock domains CLK1 and CLK2. Multi-cycle paths can be specified in DC as follows:

```
set_multicycle_path 2 -setup -from CLK1 -to CLK2
```

Note that originally the hold check was between edges 20 and 10 and would never fail. But now the hold check also becomes more restrictive.

**Scenario 24**

You have a sub-block in your design where you wish to swap all instances of a particular flip flop with flip flops with a higher drive. What is the best way to achieve this?

**Solution**

The change_link command can be used to change the reference of instances in designs to a different library cell or design. The interfaces of the new reference must be same. For example, to change the reference of an instance count_reg from FD1 to FD1P in the library libA, the following command can be used:

```
change_link count_reg libA/FD1P
```

This is a very common scenario. Most often, one might need to swap a buffer, a particular cell or maybe a pad cell. In each of these cases, the change_link command is very useful. If you are familiar with the capabilities of the cells in the technology library, then change_link command can be used very effectively to address issues such as, "If DC used the X cell in the Z path instead of the Y cell, then it should be able to meet the timing requirements."

**Scenario 25**

How does one query the technology library to find the type of delay model used for the library?

**Solution**

The get_attribute command can be used to query the technology library for the delay model used as shown below:

```
get_attribute <library_name> delay_model
```

**Scenario 26**

You are using the translate command to convert your design from one ASIC vendor technology library to another. You find that the translation process results in a design that does not meet the timing constraints.

**Solution**

The translate command does not guarantee meeting timing or area constraints after translation. In fact, no optimization occurs during the translation process. The translate command consists of two steps. First, DC performs one to one mapping of cells. For example, a 2-input AND in the netlist is replaced by a 2 input AND from the target_library. Those cells which do not have a one to one match are converted to gtech (from the Synopsys generic technology library gtech.db) logic. Then, the translate command degenerates the gtech logic and executes another mapping step.

**Scenario 27**

You are compiling a VHDL code template shown below. You find that the compile takes a long time. What might be the reasons?

```
if RESET = '1' then
  ESTORE  <= conv_unsigned(0, 320);

  elsif (CLK'event and CLK='1') then

   case WRCNTI is
    when "000000"  => ESTORE(7 downto 0) <= DATA_IN;
    when "000001"  => ESTORE(15 downto 8) <= DATA_IN;
    when "000010"  => ESTORE(23 downto 16) <= DATA_IN;
    when "000011"  => ESTORE(31 downto 24) <= DATA_IN;
    when "000100"  => ESTORE(39 downto 32) <= DATA_IN;
    when "000101"  => ESTORE(47 downto 40) <= DATA_IN;
    when "000110"  => ESTORE(55 downto 48) <= DATA_IN;
    when "000111"  => ESTORE(63 downto 56) <= DATA_IN;
    when "001000"  => ESTORE(71 downto 64) <= DATA_IN;
    when "001001"  => ESTORE(79 downto 72) <= DATA_IN;
    when "001010"  => ESTORE(87 downto 80) <= DATA_IN;
    when "001011"  => ESTORE(95 downto 88) <= DATA_IN;
    when "001100"  => ESTORE(103 downto 96) <= DATA_IN;
    when "001101"  => ESTORE(111 downto 104) <= DATA_IN;
    when "001110"  => ESTORE(119 downto 112) <= DATA_IN;
    when "001111"  => ESTORE(127 downto 120) <= DATA_IN;
```

```
      when "010000"  => ESTORE(135 downto 128) <= DATA_IN;
      when "010001"  => ESTORE(143 downto 136) <= DATA_IN;
      when "010010"  => ESTORE(151 downto 144) <= DATA_IN;
      when "010011"  => ESTORE(159 downto 152) <= DATA_IN;
      when "010100"  => ESTORE(167 downto 160) <= DATA_IN;
      when "010101"  => ESTORE(175 downto 168) <= DATA_IN;
      when "010110"  => ESTORE(183 downto 176) <= DATA_IN;
      when "010111"  => ESTORE(191 downto 184) <= DATA_IN;
      when "011000"  => ESTORE(199 downto 192) <= DATA_IN;
      when "011001"  => ESTORE(207 downto 200) <= DATA_IN;
      when "011010"  => ESTORE(215 downto 208) <= DATA_IN;
      when "011011"  => ESTORE(223 downto 216) <= DATA_IN;
      when "011100"  => ESTORE(231 downto 224) <= DATA_IN;
      when "011101"  => ESTORE(239 downto 232) <= DATA_IN;
      when "011110"  => ESTORE(247 downto 240) <= DATA_IN;
      when "011111"  => ESTORE(255 downto 248) <= DATA_IN;
      when "100000"  => ESTORE(263 downto 256) <= DATA_IN;
      when "100001"  => ESTORE(271 downto 264) <= DATA_IN;
      when "100010"  => ESTORE(279 downto 272) <= DATA_IN;
      when "100011"  => ESTORE(287 downto 280) <= DATA_IN;
      when "100100"  => ESTORE(295 downto 288) <= DATA_IN;
      when "100101"  => ESTORE(303 downto 296) <= DATA_IN;
      when "100110"  => ESTORE(311 downto 304) <= DATA_IN;
      when "100111"  => ESTORE(319 downto 312) <= DATA_IN;
      when others => ESTORE <= conv_unsigned(0, 320);
    end case;

  end if;

    end process STORAGE;
```

**Solution**

ESTORE is a 320 bit bus, so in reality what you are trying to synthesize here is 320 flops and 320 three i/p muxes and a 6/64L decoder to control the select line of the multiplexers! Hence, the long compile time.

**Scenario 28**

In a script generated from characterize, how does one distinguish between set_load -fanout_number, set_fanout_load and set_load -pin_load.

**Solution**

The pin_load is the actual pin capacitance which the output port BE[7] is driving. This is derived from the capacitance attribute on the input pins in the library.

The set_load -fanout_number specifies the number of cells the output port of the design is driving. This information is used to calculate the net load from the wire-load model specified.

The set_fanout_load specifies the summation of the fanout_load of the pins driven by this output port. The fanout_load information is used with the max_fanout constraint on your output pins or on your design. Here's a simple example

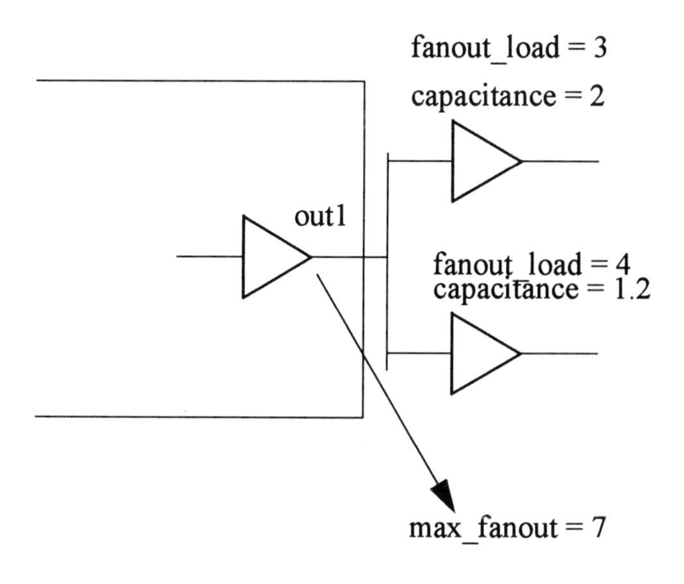


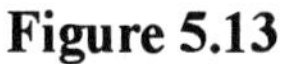
**Figure 5.13**

Results of characterize

```
set_load -pin_load {2 + 1.2} find(port, "out1")
set_load -fanout_number 2 find(port, "out1") /* Because it fans out to two places */
set_fanout_load {3 + 4} find(port, "out1")
```

**Scenario 29**

What is set_connection_class?

**Solution**

Connection rules define how DC connects components during the synthesis process. The connection class attribute is a classification label for pins and ports. Only, loads and drivers with the same connection class can be connected. In other words, loads and drivers with different connection classes cannot be connected. For example, if a driver pin has a connection class X and its loads also have connection class X, the network is considered valid. A network that has a driver pin with connection class X connected to a load with connection class Y is an invalid network. Connection rules are considered as design rules during optimization. DC tries to produce designs that meet all design rules.

**Scenario 30**

You are creating your own designware modules. Is there a way to attach an attribute to all my DesignWare components so that one can easily find them after compile.

**Solution**

This can be done by using an embedded **dc_shell** script in your implementation for that DesignWare component. Shown below is an example with an embedded dc_shell script.

```
architecture workable of test is
-- pragma dc_script_begin
-- set_attribute current_design hope true -type boolean
-- pragma dc_script_end
begin
  SUM <= A or B;
...
...
```

The following message will be issued during compile indicating that this attribute has been added to your DesignWare component

```
Information: Creating new attribute 'hope' on design 'test_6'. (UID-96)
Information: Read implementation 'workable' for synthetic design 'test_6'
```

**Scenario 31**

The VHDL code in your design causes a warning about "comparisons" to don't cares being treated as always being false (HDL-170). The code looks like this:

```
    a <= (b = "10--01--");
```

One approach is to change the code as follows:

```
a <= (b(7 downto 6) = "10" and b(3 downto 2) = "01");
```

Is there a better way to do this? The code above is easy to change, but some of the comparisons are very long and complicated. For example,

```
a <= (b = "1110----") OR (b= "1-11--00") OR (b = "--0111-0");
```

Would have to be changed to

```
a <= (b(7 downto 4) = "1110") OR ((b(7) = '1') AND (b(5 downto 4) = "11") AND
          (b(1 downto 0) = "00")) OR (b(5 downto 2) = "0111") AND (b(0) = '0'));
```

which is very hard to understand.

**Solution**

There is no way to modify the code, other than the above mentioned approach. In Verilog you could use casex to easily check for dont_cares, but in VHDL, each bit has to be specified.

**Scenario 32**

Does a synthesized netlist necessarily imply that the design will not see glitches during simulation?

**Solution**

Synopsys guarantees that if you are meeting timing constraints (using the static timing analyzer, *DesignTime*), there will be no glitches within the time frame consisting of the setup and hold time on any particular flipflop or latch. It does this using static timing analysis, and by setting proper timing constraints. For example, if the clock period is 30 ns and the setup time and hold time requirement is 1.5 ns, the tool guarantees that all paths that end at that particular flipflop will have their signals stable at time 28.5, and continue to be stable until time 31.5. For flipflops and latches this is adequate, any glitches outside this time frame will have no effect.

**Scenario 33**

How can I check if my technology library supports automatic pad synthesis. What commands can be used to list all the pad cells in the library.

**Solution**

The pad_cell attribute is used to identify a cell as an I/O pad cell. DC filters pad cells out of the normal core optimization and treats them differently during technology translation. Also, each cell that is identified as an I/O pad must have a pin that represents the pad. Such pins are identified by the is_pad attribute. Additionally the direction attribute indicates whether the cell is an input, output, or bidirectional pad. The following command can be used to find all the I/O pad cells in the lsi_10k library:

```
filter find(cell, "lsi_10k/*") "@pad_cell == true"
```

If this command returns a list of pad cells, your library supports automatic pad synthesis and the insert_pads command can be used to insert pads in your design.

**Scenario 34**

When characterizing a sub-design you get a negative value specified for set_output_delay for one of my ports. Why?

```
...
...
set_output_delay -0.4 -min -clock "CLK" find(port,"n286")
...
...
```

**Solution**

The syntax for the set_output_delay command is:

```
set_output_delay <value> -clock <clk> <other_options> <ports>
```

where <value> is the amount of time the specified port signal must be available BEFORE the active <clk> edge. For hold checks the data must be valid after the active clock edge. Hence, the negative value in the characterized script with the min option.

**Scenario 35**

The report_cell command shows a cell with a removable attribute. What does this attribute indicate?

**Solution**

The removable attribute indicates that Library Compiler knows what function the cell performs, so the cell can be replaced by a combination of other cells during synthesis and optimization. This attribute is usually attached to relatively complex cells, like adder cells, for example. These cells cannot be inferred by DC.

**Scenario 36**

When executing the report_lib command in dc_shell the output does not scroll through and displays the report a page at a time. Is there a way to turn off this feature?

**Solution**

This can be done by setting the variable enable_page_mode to false. It is set to true by default. When true, long reports are displayed one page at a time.

```
enable_page_mode = false
```

**Scenario 37**

You are using a library which contains bidirectional cells. Is it possible to find out if these cells have internal pull-ups or pull-downs on their bidirectional pin.

**Solution**

The get_attribute command can be used to get this information. For example, if you want to find out if the pin IO of the cell BD8STU of the tc160g library has a pull-up/pull-down associated with it, then the following command can be used:

```
dc_shell> get_attribute "tc160g/BD8STU/IO" driver_type
Performing get_attribute on port 'IO'.
{0}
```

The following table shows the driver_type attribute associated with the integer output of the get_attribute command

| Integer | Driver_type |
|---|---|
| 0 | pull_up |
| 1 | pull_down |
| 2 | open_drain |

**Scenario 38**

Your design has muxed inputs to some flops with the Q output being one of the inputs to the mux. This results in a path from clk to Q to mux input A to D input of the same flop. The time taken is 1.16ns. The hold time of the flop is 0.4ns so obviously a hold violation does not exist. However, if you add uncertainty of 1ns the clock, report_timing -delay min reports a hold violation ( 1.0 + 0.4 > 1.16 ). Why?

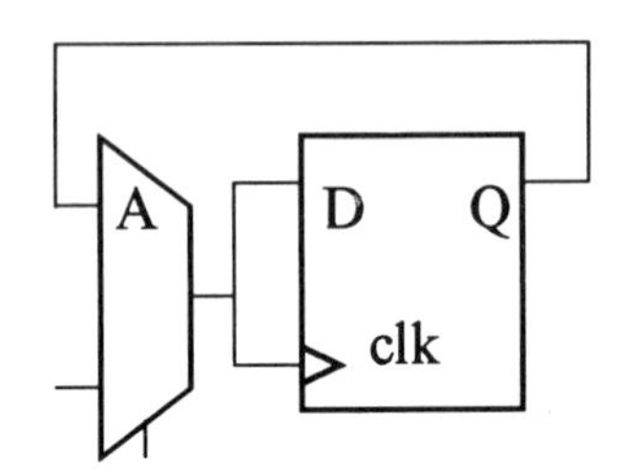


**Figure 5.14 Design with Feedback Loop**

**Solution**

One of the following approaches can be used to address this scenario. To prevent this behavior one can set the dc_shell variable, set timing_self_loops_no_skew to TRUE. Alternatively, one can use the dc_shell script shown below.

```
filter find(cell,"*" -hier) "@is_sequential==true"
foreach (item,dc_shell_status) {
set_false_path -from item -to item -hold
}
```

## Recommended Further Readings

1. Design Compiler Family Reference Manual
2. Technology Translation Application Note.
3. Synopsys Newsletter *"Impact"* Support Center Q&A.

# CHAPTER 6 *Links to Layout*

Major advances in semiconductor technologies have made possible ICs under 0.3 micron technology. As a result of these shrinking geometries, net delay is fast becoming a major component of path delays. Hence, it is essential that synthesis tools take into consideration information provided by tools used downstream in the design flow like floor-planners and place and route tools. This chapter discusses the links from and to DC and backend tools like floorplanners and place and route tools.

After the synthesis process, the next step involves floorplanning, followed by place and route. DC provides a mechanism for incorporating more accurate real-world delays generated after both floorplanning and place and route. In other words, it is possible to reoptimize the design in DC to account for the delays after place and route. This chapter provides a comprehensive discussion of the different links to layout methodologies and techniques involving the Synopsys *Floorplan Manager*.

## 6.1 Motivation for Links to Layout

CAD tools like those meant for schematic capture, logic synthesis, simulation and timing analysis are often referred to as front end tools. Similarly, tools that deal with the "physical" world like floorplanners and place and route tools are referred to as backend tools. With the shrinking of process geometries, the delays incurred by the switching of transistors become smaller. On the other hand, delays due to physical characteristics (R, C) of wires connecting the transistors become larger. Front end tools like DC do not take into consideration "physical" information like placement when optimizing the design. Further, the wire-load models specified in the Synopsys technology library are often based on statistical data. This data is extremely design and process technology dependent. The convergence between DC synthesis timing and post-layout timing depends largely on the accuracy of these wire-load models and the actual placement and routing. In-accuracies in wire-load models can lead to synthesized designs which are un-routable or don't meet timing requirements after routing.

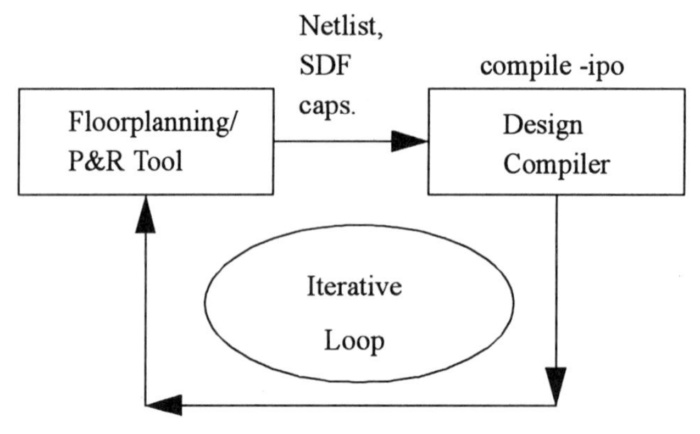


**Figure 6.1 Design Flow using Design Compiler**

In general, designs are partitioned based on the functionality rather than by physical concepts like connectivity. However, placement and route is often performed on a flat netlist. In such cases, the functional hierarchy looses its significance during place and route. But, the actual placement of the different cells of a functional hierarchy determines the delays in the design. Thus, it is increasingly important that physical characteristics be accounted for during the logic synthesis phase. In other words, it is imperative that there be a mechanism for transfer of information from and to synthesis tools and backend tools. The design flow using DC is shown in Figure 6.1. Note that this flow does not include back-annotation of cluster (placement) information. The Synopsys *Floorplan Manager*, which is fully integrated into DC, facilitates the links from and to DC and backend tools. In other words, it helps DC understand back-annotated information like clustering, timing, and parasitics.

## 6.2 Floorplanning

A Floorplanning tool helps define the dimensions of the chip layout and place modules of the design in specific regions. Some Floorplanning tools also have an option to do automatic floorplanning. At this stage of the design flow, this is a very rough estimate of the actual placement. Also, no actual routing is done at this stage, so there is no guarantee that the floorplanned design can be effectively routed by the P&R tool. While routing a design is often a compute intensive task, placement is a relatively fast process. Floorplanning tools can provide estimated parasitic capacitance and cluster information that can be fed back to the Floorplan Manager. The Synopsys Floorplan Manager ensures that this information is utilized during subsequent synthesis. There a two basic advantages to doing this. Firstly, this improves the probability of the design meeting timing requirements after final place and route. Secondly, this eliminates the need for compute intensive routing iterations. Further, placement tools are relatively easy to use and are conceived and developed with the designer in mind. In other words, the intent is to assist designers to perform placement upfront and analyze the design using estimated loads before performing complete routing on a design.

## 6.3 Link to Layout Flow Using FloorPlan Manager

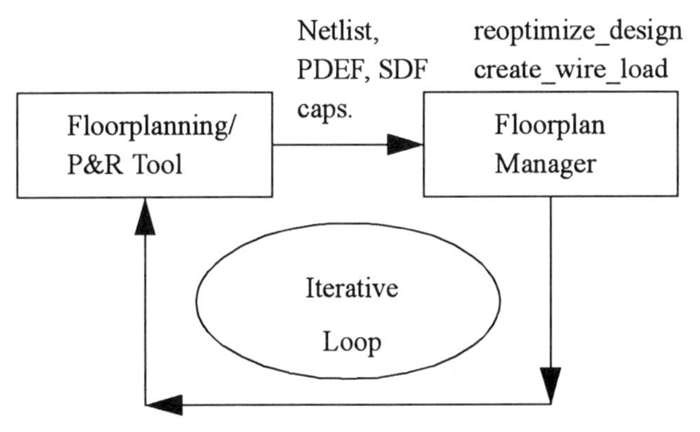


**Figure 6.2 Design Flow using Floorplan Manager**

Once floorplanning is completed (either automatically or manually by the user), the floorplanner can write out estimated parasitics and cluster information. This data is then provided to the Floorplan Manager as shown in Figure 6.2. After place and route, the parasitics generated are more accurate since they are based on actual routed data. The *Floorplan Manager* helps ensure an effective link between DC and floorplanning and place and route tools. Interchange of information between these tools is accomplished using Standard Delay Format(SDF), Physical Data Exchange Format (PDEF), and Resistance/Load script. The following steps describe the links to layout flow using the Synopsys Floorplan Manager.

1. Synthesize your design to gates using the statistical wire-load models provided in the technology library. This is likely to be inaccurate since wire-loads are very much design and layout dependent. At this stage, one can write out timing constraint information in SDF from DC for timing driven floorplanning/place & route tools.

2. Perform floorplaning, followed by placement of the entire design. Pre-placement, grouping and regions are performed at this stage. Generate the cluster information (PDEF), estimated parasitics (RC) and delay values (SDF).
3. Back-annotate physical information from placement in the Floorplan Manager. Create custom wire-load models for your design based on the back-annotated data. This wire-load can be based on "physical" grouping performed during step 2.
4. Perform static timing analysis on the design after back-annotation. If the timing violations are relatively large, re-synthesize design beginning with source HDL and new wire-load models generated in step 3. On the other hand if the timing violations are small, reoptimize the design using the reoptimize_design command with the -post_layout option. This option has far greater flexibility in making changes to the design than the ipo option since at this stage actual routing has not been performed. Generate netlist after re-synthesis, new PDEF and SDF file for further floorplanning.
5. Repeat steps 2, 3 and 4 until the design meets timing with sufficient slack. Then proceed to perform complete routing and generate back-annotation data in SDF, PDEF and parasitics (set_load/seat_resistance)
6. At this point, the timing violations, if any, should be fixed by performing in place optimization on the design using the reoptimize_design -ipo command as discussed in section 6.5.

### 6.3.1 Standard Delay Format (SDF)

SDF is an industry wide standard for transferring timing information among different EDA vendors. Synthesis supports a subset of the available SDF constructs and not all the constructs. DC can forward annotate design timing constraints to external Floorplanning and Placement and Route tools through the SDF format. This is possible by executing the write_constraints command. There are options available with this command to control the number of paths, max/min paths, upper limit on timing slack written out and so on. DC can also back-annotate net/cell delays after floorplanning and placement and routing using the read_timing command.

**Example 6.1 DC Timing Report and corresponding SDF from DC**

```
Startpoint: SEEN_TRAILING_reg
        (rising edge-triggered flip-flop clocked by CLK)
Endpoint: COUNT_reg[1]
        (rising edge-triggered flip-flop clocked by CLK)
Path Group: CLK
Path Type: max
```

```
Point                          Incur   Path
-----------------------------------------------------------
clock CLK (rise edge)          0.00    0.00
clock network delay (ideal)    0.00    0.00
SEEN_TRAILING_reg/CP (FD1)     0.00    0.00 r
SEEN_TRAILING_reg/Q (FD1)      1.47    1.47 f
U114/Z (IVP)                   0.49    1.97 r
U93/Z (NR2)                    0.49    2.45 f
U92/Z (OR3)                    1.38    3.83 f
U95/Z (NR2)                    1.84    5.68 r
U111/Z (ND2)                   0.22    5.89 f
U98/Z (AO2)                    1.08    6.98 r
COUNT_reg[1]/D (FD1)           0.00    6.98 r
data arrival time                      6.98

clock CLK (rise edge)          4.00    4.00
clock network delay (ideal)    0.00    4.00
COUNT_reg[1]/CP (FD1)          0.00    4.00 r
library setup time             -0.80   3.20
data required time                     3.20
-----------------------------------------------------------
data required time                     3.20
data arrival time                      -6.98
-----------------------------------------------------------
slack (VIOLATED)                       -3.78
```

## SDF Output from write_constraints

```
(CELL
 (CELLTYPE "COUNT_SEQ_VHDL")
  (INSTANCE)
  (TIMINGCHECK
   (PATHCONSTRAINT SEEN_TRAILING_reg/CP SEEN_TRAILING_reg/Q U114/A
U114/Z U93/A U93/Z U92/B U92/Z U95/A U95/Z U111/A U111/Z U98/A U98/Z
COUNT_reg\[1\]/D (3.200:3.200:3.200) )


  )
)
```

)

Example 6.1 shows a max timing violation path in a design and the corresponding timing constraint in the SDF file. This SDF file can be forward-annotated to back-end tools. The floorplanner and/or the P&R tool on reading the SDF file interprets that the required time for the path starting SEEN_TRAILING_reg/CP and ending at COUNT_reg\[1\]/D is 3.2. This SDF timing constraint file can be used as input to the floorplanning tool along with the design netlist.

### Example 6.2 SDF File after Floorplanning

```
(DELAYFILE
(SDFVERSION "OVI 1.0")
 (DESIGN "top")
 (DATE "Mar-16-96")
 (VENDOR "Compass")
 (PROGRAM "HDL Asst")
 (VERSION "1.0")
 (DIVIDER .)
 (VOLTAGE )
 (PROCESS "")
 (TEMPERATURE )
 (TIMESCALE 1 ns)

 (CELL (CELLTYPE "an05d2")
  (INSTANCE U11)
  (DELAY (ABSOLUTE
   (IOPATH A4 Z (1.174:1.174:1.174) (.862:.862:.862))
   (IOPATH A2 Z (1.168:1.168:1.168) (.772:.772:.772))
   (IOPATH A5 Z (1.219:1.219:1.219) (.845:.845:.845))
   (IOPATH A3 Z (1.252:1.252:1.252) (.801:.801:.801))
   (IOPATH A1 Z (1.309:1.309:1.309) (.787:.787:.787))
 ))
 )
..
..
..

 (CELL (CELLTYPE "dfntnb")
  (INSTANCE I1.int_count_reg_3)
  (DELAY (ABSOLUTE
```

```
    (IOPATH CP Q (1.144:1.144:1.144) (.797:.797:.797))
    (IOPATH CP QN (.714:.714:.714) (.794:.794:.794))
   ))
   (TIMINGCHECK
    (SETUP D (posedge CP) (.280))
    (HOLD D (posedge CP) (.000))
    (WIDTH  (posedge CP) (.300))
    (WIDTH  (negedge CP) (.350))
   )
  )
..
..
..

  (CELL (CELLTYPE "top")
  (INSTANCE )
  (DELAY (ABSOLUTE
    (INTERCONNECT  U11.Z I2.int_count_reg_3.CP (.050:.050:.050) (.052:.052:.052))
    (INTERCONNECT  U11.Z I2.int_count_reg_1.CP (.078:.078:.078) (.081:.081:.081))
    (INTERCONNECT  U11.Z I2.int_count_reg_0.CP (.089:.089:.089) (.092:.092:.092))
    (INTERCONNECT  U12.Z TC_reg.D (.006:.006:.006) (.006:.006:.006))
    (INTERCONNECT  I1.int_count_reg_3.Q U11.A1 (.056:.056:.056) (.053:.053:.053))
    (INTERCONNECT  I1.int_count_reg_2.Q U11.A4 (.046:.046:.046) (.047:.047:.047))
    (INTERCONNECT  I1.int_count_reg_2.Q I1.U8.A2 (.002:.002:.002) (.002:.002:.002))
    (INTERCONNECT  I1.int_count_reg_1.Q I1.U7.A1 (.017:.017:.017) (.017:.017:.017))
    (INTERCONNECT  I1.int_count_reg_1.Q I1.U9.A1 (.017:.017:.017) (.017:.017:.017))
    (INTERCONNECT  I1.int_count_reg_0.Q U11.A3 (.019:.019:.019) (.019:.019:.019))
    (INTERCONNECT  I2.int_count_reg_3.Q U12.A4 (.042:.042:.042) (.040:.040:.040))
    (INTERCONNECT  I2.int_count_reg_2.Q U12.A3 (.079:.079:.079) (.075:.075:.075))
    (INTERCONNECT  I2.int_count_reg_2.Q I2.U8.A2 (.016:.016:.016) (.010:.010:.010))
    (INTERCONNECT  I2.int_count_reg_2.Q I2.U10.A1 (.021:.021:.021) (.015:.015:.015))
    (INTERCONNECT  I2.int_count_reg_1.Q U12.A2 (.039:.039:.039) (.044:.044:.044))
    (INTERCONNECT  I2.int_count_reg_1.Q I2.U9.A1 (.032:.032:.032) (.037:.037:.037))
    (INTERCONNECT  I2.int_count_reg_0.Q U12.A1 (.006:.006:.006) (.006:.006:.006))
    (INTERCONNECT  I2.int_count_reg_0.Q I2.U7.A2 (.007:.007:.007) (.007:.007:.007))
  ))
  )
)
```

Example 6.2 shows an SDF file written out from a commercial floorplanning tool. The SDF file provides information about the estimated cell and net delay from floorplanning evaluations. There are two delays defined in the SDF file IOPATH and INTERCONNECT delay. The IOPATH delay is the cell intrinsic delay and the INTERCONNECT delay is the net connect delay. The Floorplanner tool may choose to lump the net transition (load) delay either with the IOPATH or the INTERCONNECT delay. Also the setup and hold check information is provided for the sequential elements.

### 6.3.2 Physical Data Exchange Format(PDEF)

PDEF is the industry standard initiated by Synopsys for exchanging physical cluster information between DC and backend tools. The PDEF file contains physical cluster information generated from floorplanning or from DC. It is possible to group cells in DC using the group command with the -soft option. This information can be forward-annotated to the floorplanner by writing out PDEF from DC using the Synopsys Floorplan Manager.

#### Example 6.3 PDEF File

```
(CLUSTERFILE
  (PDEFVERSION "1.0")
  (DESIGN "top")
  (DATE "Sat Mar 16 11:33:10 1996")
  (VENDOR "COMPASS")
  (PROGRAM "ChipPlanner-GA")
  (VERSION "v8r4.9.0")
  (DIVIDER /)
  (CLUSTER (NAME "I2")
    (UTILIZATION .4328)
    (MAX_UTILIZATION 100.0000)
    (CELL (NAME I2/int_count_reg_2))
    (CELL (NAME I2/int_count_reg_3))
    (CELL (NAME I2/U8))
    (CELL (NAME I2/U11))
    (CELL (NAME I2/U10))
    (CELL (NAME I2/U7))
    (CELL (NAME I2/int_count_reg_1))
    (CELL (NAME I2/int_count_reg_0))
    (CELL (NAME I2/U9))
  )
```

```
  (CLUSTER (NAME "top_ga")
    (UTILIZATION .4379)
    (MAX_UTILIZATION 100.0000)
    (CELL (NAME U11))
    (CELL (NAME TC_reg))
    (CELL (NAME U12))
  )
  (CLUSTER (NAME "I1")
    (UTILIZATION .4361)
    (MAX_UTILIZATION 100.0000)
    (CELL (NAME I1/int_count_reg_1))
    (CELL (NAME I1/int_count_reg_0))
    (CELL (NAME I1/U9))
    (CELL (NAME I1/U7))
    (CELL (NAME I1/U8))
    (CELL (NAME I1/U10))
    (CELL (NAME I1/int_count_reg_2))
    (CELL (NAME I1/int_count_reg_3))
    (CELL (NAME I1/U11))
  )
)
```

Example 6.3 shows a PDEF file written out by the Floorplanner from Compass Design Automation. The PDEF file shows three clusters for this design: I1, I2 and top_ga. Also, information provided in the PDEF file is the actual cell instance names in each of the clusters.

### 6.3.3 Parasitics and Resistance

#### Example 6.4 Net Parasitics and Net Resistance

```
auto_link_disable = true
set_load 1.5801 "clk"
set_resistance .3863 "clk"
set_load .5782 "TC"
set_resistance .1258 "TC"
set_load .8826 "temp1_3"
set_resistance .2183 "temp1_3"
set_load .6500 "temp2_1"
```

```
set_resistance .1569 "temp2_1"
set_load .6730 "temp1_2"
set_resistance .1633 "temp1_2"
set_load .8976 "gated_clk"
set_resistance .2218 "gated_clk"
```

Example 6.4 shows net resistance and net parasitic file written out from a commercial Florrplanning tool. In other words, the tool provides estimated resistance and capacitance values for each net in the design. This information can be back-annotated to DC to generate new more accurate wire load models as discussed in section 6.4 or to perform in place optimization.

## 6.4 Basic Links to Layout Commands

The information written out by the Floorplanner is read into DC. Then, one must perform static timing analysis and if necessary to re-optimize the design. Listed below are the basic commands that help transfer of information to and from DC.

| | |
|---|---|
| read_clusters/write_cluster | Used to read/write PDEF |
| read_timing/ write_timing | To read/write SDF |
| include <file_name> | To read in the parasitics and net resistances. |
| write_constraints | To write timing constraint information in SDF for timing driven floorplanning and P&R. |
| create_wire_load | Used to create custom wire-load models after back-annotation of estimated loads. |

## 6.5 Creating Wire Load Models After Back-Annotation

Wire-load models specified in the technology library are critical to achieving a design that is both routable and meets timing constraints. The wire-load model used during synthesis determines the estimated capacitance, area and the length of the nets in a design. The wire load model is chosen using one of the following mechanisms during synthesis.

1. Area-based selection from technology library.
2. The designer explicitly specifies the desired wire load model.
3. Default wire load specified is in technology library.

In general, wire load models are entirely dependent on the design and the process technology. Thus, it is extremely difficult to have an accurate wire-load model the first time a design is being targeted to a certain technology. As designs move into the sub-micron geometries, wire delays contribute significantly to the overall delay. Hence, accurate wire load models are critical to achieving designs that meet timing constraints after place and route. The Floorplan Manager helps generate wire-load models based on delay estimates from placement information. In other words, one can synthesize HDL to gates and perform floorplanning early in the design cycle. Most Floorplanning tools can generate PDEF, SDF and net parasitics and resistance based on estimated delays. This information can be read into Synopsys and used to create a custom wire-load model - models which are design and technology specific. This information can then be used during synthesis of HDL to gates.

### 6.5.1 Steps for Generating Wire Load Models

1. Read in design (netlist, db).
2. Include set_load script after floorplanning.
3. create_wire_load

The set_load script specifies the load on every net in the design. Most floorplanning and place and route tools are capable of generating this information. The create_wire_load command be used with the -hierarchy option to generate wire load models for each sub-block in the design.

### Example 6.5 Custom Wire-load Model

```
/* Synopsys wire load models for design top */
/* format: fanout_length(fanout, length, average_cap, std_dev, points); */
wire_load("compass_wl") {
 resistance : 0;
 capacitance : 1;
 area : 0;
 slope : 0.48;
 fanout_length( 1, 0.26, 0.26, 0.059, 11);
 fanout_length( 2, 0.26, 0.26, 0.16, 2);
 fanout_length( 3, 0.62, 0.62, 0.066, 6);
 fanout_length( 4, 1.1, 1.1, 0.2, 1);
}
```

Example 6.5 shows a custom wire-load model generated from Compass Design Automation's floorplan tool. This custom wire load model can be included in your technology library using the update_lib command and then selected appropriately using the set_wire_load command. These estimated wire-loads are not as accurate as actual wire-loads based on place and route generated wire loads. However, these wire-loads do minimize the number of elaborate, time consuming place and route cycles that one has to go through before arriving at an accurate wire model.

## 6.6 Re-Optimizing Designs After P&R

After placement and routing the design has to be reoptimized after back-annotation of delays generated from the layout. If after back-annotation, the design does not meet timing constraints by a large margin, one can re-synthesize the design beginning with the source HDL and more accurate wire-load. On the other hand, if the timing constraints after a back-annotation are violated by a relatively small margin, one can re-optimize the existing netlist. This is performed using the reoptimize_design command. The reoptimize_design command (used without any options) is similar to compile -incremental. However, if physical clustering information is back-annotated, the automatic wire load selection during reoptimize_design is directed on the basis of physical grouping, not logical grouping as in the case of the compile command. The compile -incremental command uses existing gates as a starting point for the mapping process. Mapping optimizations involving existing gates are accepted only if they improve the circuit speed, porosity and area.

The reoptimize_design command is used with the -post_layout option after placement and not after routing. This is because this command is capable of making large-scale changes in the design which might make it impossible to perform any incremental P&R. However, the reoptimize_design command is used with the -ipo option after routing. This ensures that back-annotated timing information (SDF) and placement information (PDEF) are considered during the reoptimization process. It is also possible to perform in-place optimization using the compile -ipo command. This does not however, consider placement information (PDEF) during the optimization process.

### 6.6.1 In-Place Optimization

In-place optimization aims to meet the timing for the critical path by swapping cells. In sub-critical paths, the tool swaps cells to improve area, provided the timing constraints are not violated. This swap down helps to bring down the power consumption of the ASIC indirectly. After synthesis and simulation, a preliminary place and route provides information of realistic delay and load values. Feeding back this information into DC to reoptimize the design without modifying its existing structure, but by merely swapping cells in the design with higher or lower drive cells from the technology library, is the essence of in-place optimization. Several factors have a significant impact on successful in-place optimization. DC has a number of IPO specific compile variables. One can list the compile variables using the following command:

list -var compile

When back-annotating load values into DC, the new constraint violations are dependent entirely on the accuracy of the wire load models used during pre-layout synthesis. How were the real time delays observed after the place & route phase estimated during the initial synthesis? In other words, were the violations after place and route too large? Just how accurate the pre-synthesis "guesstimates" were, determine the constraint violation in DC after backannotation. Design houses use several different strategies to ensure that the timing violations after place and route are not too large. Some over-estimate capacitance values in the library by a certain value for the pre-layout synthesis phase. This value is derived from past experience and by iterative steps. In other words, all capacitance values in the library are over-estimated by a certain factor to compensate for wire load. Others prefer to set a uniform load value on all nets in the design for pre-layout synthesis in addition to any default values in the library. Having an accurate wireload model and a non-linear delay model technology library will usually provide better results. What follows is a discussion of the issues affecting in-place optimization, and finally, the steps to be followed for in-place optimization.

The most crucial factor in the IPO process are the values and attributes on cells in the technology library. All cells that can be swapped must have footprints attached to them. DC swaps only those cells that have the same footprint. This behavior can be controlled by dc_shell compile variables. The more the number of cells with the same footprint, the greater the choice of cells available to DC. One must ensure that swappable cells do not have dont_touch or dont_use attributes set on them in the library.

To meet the timing after backannotation, DC must have alternate faster (lower resistance, higher drive) cells to replace existing cells in the critical path of the design. In general, synthesis is most effective when the technology library provides a greater choice of library cells. For in-place optimization, having a wide choice of low and high drive cells with the same functionality aids the *swap-up* and *swap-down* process significantly. The low drive cells must preferably have lower area. If the areas of the cells are identical, you must set the variable compile_ignore_area_during_inplace_opt to true.

During in-place optimization, ensure that a max_transition constraint is specified in the top level constraint script in the design or via a default_max_transition attribute in the technology library. The max fanout of the output pin considers only the fanout_load of all the pins that are driven by the output pin, but not the load on the nets. If your backannotation methodology involves a set_load script, a max transition constraint on the design is essential. Max_transition constraints ensure that the product of the capacitive load on a net (set by the set_load command) and the resistance of the output pin driving the net does not exceed the max_transition value. After running a constraint report, if there are certain specific paths you might be more concerned about, assign a higher weightage to them using the group_path command. Also use the critical_path option, if necessary.

compile_default_critical_range = 2.0

If there are specific paths in the design (the clock network, for example), whose cells you do not wish to swap during IPO, set a dont_touch_network attribute on them. Similarly, ungrouping the different levels of hierarchy after backannotating the loads (DC maintains the load values after ungroup) has a significant impact on the quality of results, that is, improvement with regard to meeting constraints. However, if you wish to write out a netlist after ungroup (and IPO), this might make it impossible to use the existing testbench for simulation, since the instance names assigned will be names showing the hierarchy. An alternate approach might be to use a perl script which identifies the swapped cells and re-generates a new netlist with the swapped instances. Alternatively, you can regenerate the netlist by executing a dc_shell script which uses the change_link command to link swapped instances to their new references.

## 6.6.2 In place Optimization Steps

The in-place optimization steps are as follows:

1. Read in compiled db.
2. Include top level constraint file.
3. Include script with back annotation information, that is, a dc_shell set_load script with load information for all nets in the design and an SDF file.
4. Write out the hierarchical db file. Use this for future re-runs. This helps, since the set_load script is usually long and takes a significant time to complete, depending on the size of the design.
5. Read the technology library (db file) into DC.
6. Include the area script to set area values for the different cells. This script essentially specifies lower area values for low drive cells. For this script, one must have a list of all the low drive cells in the library. This information will have to be obtained from the ASIC vendor, unless the vendor follows a particular naming convention for low drive cells.
7. Set max_transition for the entire hierarchical design.
8. Set the dont_touch attribute on any blocks in the design which you wish to remain unchanged during IPO.
9. Set current_design to the top level block, and ungroup all the levels below it. The dont touched blocks will not be ungrouped.

10. Place a `dont_touch` attribute on the clock network. (dont_touch_network <clock_port>)
11. Specify the file name for the list of instances swapped. reoptimize_design_changed_list_file_name = swap-cells
12. Set the area constraint. (max_area 0.0)
13. Perform in-place optimization. (reoptimize_design -in_place -map_effort high)

After performing in-place optimization a few times (that is, repeating step 13), no further swapping will be possible and the swap-cells file will show no additional swaps. DC appends the list of cells to the swap-cells file each time a compile is performed and does not overwrite the existing file.

### Example 6.6 A List of Cells Swapped During IPO

```
$ START_CHANGES_FILE
$ Synopsys, Inc.
$ Design:  core

# Report on the changes made during optimization.
$ MODIFIED_CELLS
# <cell_name> (<new_ref>) : original lib_cell (<old_ref>)
aq/U427 (invl)  :  original lib_cell (inv)
aq/U428 (nor2)  :  original lib_cell (nr2l)
aq/U429 (buff12) :  original lib_cell (buff3)
aq/U430 (nor2)  :  original lib_cell (nr2l)
aq/U431 (buff12) :  original lib_cell (buff3)
$ END_CHANGES_FILE
```

DC provides a list of instances which were swapped during IPO. If cells are added during the IPO process (adding cells is dependent on compile variables and not done by default), then this information is also added to this file.

## 6.7 Classic Scenarios

**Scenario 1**

You are doing post-route static timing analysis using SDF file generated from layout tools. Do wire load models affect the timing after back-annotation? You get different results for different wire load models.

**Solution**

If the tool that generated the SDF file lumped the transition delay in with the net delay rather than the cell delay (DC assumes that the transition time is by default included in the cell delay), and if you have not back annotated your capacitance information as well, then when *DesignTime* tries to subtract the transition delay from the net delay it must base the calculations of transition delay on the wire load model. When you read the back annotated timing, you should set the two variables if the transition time has been included in the net delay.

```
read_net_transtion = true (default is false)
read_cell_transition = false (default is true)
```

If you set the above variables, DC calculates the transition delay and subtracts it from the net delay and adds it to the cell delay. It uses the back annotated capacitances to calculate the transition delay. However, if you don't back annotate your capacitances, DC must use the capacitances in the wire load model. This might explain the differences you saw with different wire load models after back annotating the timing.

Note that in v3.3a and subsequent versions of DC, the variables read_cell_transition and read_net_transition will be obsolete. The user will instead specify where the transition delay is by using the load_delay option with read timing. Thus, instead of:

```
read_net_transition = true
read_cell_transition = false
read_timing
```

in version v3.3a and after of DC one should use:

```
read_timing -load_delay net
```

Likewise for:

```
read_net_transition = false
read_cell_transition = true
read_timing
read_timing -load_delay cell        (-load_delay cell is the default)
```

**Scenario 2**

You execute report_timing with the -net option. But this does not show the net delay component.

```
Startpoint: SEEN_ZERO_reg
        (rising edge-triggered flip-flop clocked by CLK)
```

```
Endpoint: SEEN_TRAILING_reg
        (rising edge-triggered flip-flop clocked by CLK)
Path Group: CLK
Path Type: max

Point                       Fanout    Incr      Path
----------------------------------------------------------
clock CLK (rise edge)                 0.00      0.00
clock network delay (ideal)           0.00      0.00
SEEN_ZERO_reg/CP (FD1)                0.00      0.00 r
SEEN_ZERO_reg/Q (FD1)                 3.26      3.26 r
SEEN_ZERO (net)       2               0.00      3.26 r
U127/Z (ND2)                          0.70      3.97 f
n261 (net)                  1         0.00      3.97 f
U115/Z (AO4)                          3.13      7.09 r
n285 (net)                  1         0.00      7.09 r
SEEN_TRAILING_reg/D (FD1)   0.00      7.09 r
data arrival time                                7.09

clock CLK (rise edge)                 10.00      10.00
clock network delay (ideal)           0.00       10.00
SEEN_TRAILING_reg/CP (FD1)0.00                   10.00 r
library setup tim                     -0.80      9.20
data required time                               9.20
--------------------------------------------------------------------
data required time                               9.20
data arrival time                                -7.09
--------------------------------------------------------------------
slack (MET)                                      2.11
```

**Solution**

Yes, report_timing -net shows the fanout of each net but does not show the net delay component. It is possible to find the actual net delay but it is a little complicated. This is the relevant section from the timing report,

```
...
SEEN_ZERO_reg/Q (FD1)              3.26      3.26 r
SEEN_ZERO (net)              2     0.00      3.26 r
U127/Z (ND2)                       0.70      3.97 f
```

...

For example, you can find the net delay of the net SEEN_ZERO by writing out the SDF file as shown below. The SDF file can be written out with the delay due to the loading on the net as a separate INTERCONNECT component in the SDF file.

```
write_timing -format sdf -load_delay net
```

### Relevant sections of the SDF file

```
(CELL
 (CELLTYPE "sdf")
 (INSTANCE)
 (DELAY
  (ABSOLUTE
  (INTERCONNECT SEEN_ZERO_reg/Q U127/A (1.233:1.233:1.233)
(0.442:0.442:0.442))
  (INTERCONNECT U151/Z U115/C (1.221:1.221:1.221) (0.498:0.498:0.498))
  (INTERCONNECT U127/Z U115/D (0.744:0.744:0.744) (0.462:0.462:0.462))
  (INTERCONNECT U115/Z SEEN_TRAILING_reg/D (1.412:1.412:1.412)
(0.445:0.445:0.445))
  )
 )
)

...
...
(CELL
 (CELLTYPE "FD1")
 (INSTANCE SEEN_ZERO_reg)
 (DELAY
  (ABSOLUTE
  (IOPATH CP Q (2.031:2.031:2.031) (2.553:2.553:2.553))
  (IOPATH CP QN (2.963:2.963:2.963) (2.926:2.926:2.926))
  )
 )
 (TIMINGCHECK
  (WIDTH (posedge CP) (1.500:1.500:1.500))
  (WIDTH (negedge CP) (1.500:1.500:1.500))
  (SETUP D CP (0.800:0.800:0.800))
```

```
    (HOLD D CP (0.400:0.400:0.400))
  )
)
```

The timing report lumps the total cell and net delay at the output of the source. So as seen from the timing report the delay at the output pin

SEEN_ZERO_reg/Q is 2.031(rise delay from CP -> Q) + 1.233 (INTERCONNECT SEEN_ZERO_reg/Q U127/A) = 3.264.

**Scenario 3**

How is the reoptimize_design command different from compile -in_place or compile -inc?

**Solution**

The reoptimize_design command (used without any options) is similar to compile -incremental except if physical clustering information is back-annotated, the automatic wire load selection is directed on the basis of physical grouping, not logical grouping. reoptimize_design -post_layout performs restructuring along the critical paths (only) while maintaining the rest of the design intact. Use post layout optimization after floorplanning (but not after the place-and-route) when there are a large number of timing violations because it is capable of making big changes to the design

The compile -incremental command uses existing gates as a starting point for the mapping process. Mapping optimizations involving existing gates are accepted only if they improve the circuit speed, porosity, area. Using incremental mapping guarantees that a circuit can only be improved.

**Scenario 4**

You are writing out an SDF file from Design Compiler and do not get any INTERCONNECT delay being written out even though you have specified the wire-load models.

**Solution**

The INTERCONNECT delay consists of the Connect Delay component and maybe the Load Delay. When writing out SDF file from DC, you can specify whether the Load Delay should be included in the IOPATH delay or INTERCONNECT by using the appropriate options with the write_timing command. Also if the resistance value in the wire-load model selected is 0, the Connect Delay component shall be 0. Here is an example.

```
wire_load("10x10") {
    resistance : 0 ;
    capacitance : 1 ;
    area : 0 ;
    slope : 0.311 ;
    fanout_length(1,0.53) ;
}
```

Also if tree_type in the operating conditions selected is best_case_tree, the wire resistance is taken as 0 and the Connect Delay is 0. Here is an example.

```
operating_conditions(BCCOM) {
    process : 0.6 ;
    temperature : 0 ;
    voltage : 5.25 ;
    tree_type : "best_case_tree" ;
```

If no operating conditions have been specified, the default tree_type assumed is balanced_tree.

**Scenario 5**

The PDEF file from your floorplanner specifies a max_utilization number for each of the clusters.

```
(CLUSTER (NAME "I2")
    (UTILIZATION 83.28)
    (MAX_UTILIZATION 75.0000)
    (CELL (NAME I2/int_count_reg_2))
    (CELL (NAME I2/int_count_reg_3))
)
```

Why doesn't DC optimize the design to meet the max_utilization requirement?

**Solution**

Max_utilization is not used as a constraint during synthesis by DC. This information is used only to pass PDEF from DC to backend tools. Currently DC only uses the utilization information to decide which cluster to optimize first.

**Scenario 6**

You wish to prevent high fan-in cells from being inferred during synthesis. These cause routability issues during P&R. How does one deal with this?

**Solution**

The dont_use attribute can be used on the high fan in cells prior to synthesizing the design. Here is an example which eliminates all cells with more than 6 input pins from the lsi_10k library for synthesis.

```
filter find(cell, "lsi_10k/*") "@number_of_pins >= 6"
set_dont_use dc_shell_status
```

## Recommended Further Readings

1. *Synopsys Links-to-Layout Methodology Flow Application Note*

# CHAPTER 7 FPGA Synthesis

Field Programmable Gate Arrays (FPGAs) have emerged as a quick and very effective means to generate low-cost prototypes of designs. In recent times, FPGAs have grown from a tiny market niche to a significant portion of the IC market. The complexity and speed of the FPGAs available in the market has been increasing at a rapid pace. Simultaneously, the cost per gate of FPGAs has been fast decreasing. The Synopsys *FPGA Compiler* has been developed primarily to target FPGA technology libraries. The *FPGA Compiler* is fully integrated into the Synopsys Design Compiler/Design Analyzer front end. For a user familiar with DC, *FPGA Compiler* is easy to use.

This chapter provides an overview of the methodology for targeting designs to FPGA libraries. It discusses FPGA synthesis with special focus on Xilinx 4000 family of FPGAs. However, the basic concepts are applicable to other FPGAs. We first provide a brief introduction to FPGA synthesis, assuming the reader is familiar with logic synthesis for ASICs. This is followed by a detailed description of the Xilinx 4000 (XC4000) architecture. Then, a sample Synopsys setup is provided to illustrate the environment required for *FPGA Compiler*. Finally, the entire FPGA synthesis flow is outlined. Synopsys announced FPGA Express for the PC platform at DAC 96. This chapter discusses the FPGA Compiler and not the FPGA Express.

## 7.1 FPGAs vs. ASICs

FPGAs are arrays of uncommitted elements whose interconnections can be user-programmed. Since FPGAs can be programmed and verified in a matter of a few days, they prove to be very effective at prototyping logic devices. This enables users to get equipment into production quickly, without non-recurring engineering costs.

FPGAs differ from ASICs in a number of ways. The granularity of primitive elements in FPGAs is generally much larger than in ASICs. In other words, FPGAs can provide complex architectural resources. The complexity however, limits the number of these resources on the FPGA. Further, routing resources in FPGAs are limited. Two main disadvantages of FPGAs, are their relatively low speed of operation and lower density when compared to ASICs. The amount of logic circuitry possible on a single ASIC often requires several FPGAs. This means that fewer FPGAs chips can be produced per wafer. On the other hand, FPGAs have lower prototype costs and shorter production times. Hence, most design houses target FPGAs to design low cost prototypes which can be used in efforts such as emulation. *Emulation* is a technique of using programmable hardware as a physical prototype of custom and semi-custom ICs prior to silicon fabrication for chip and sytem verification. The ASIC design of the same chip is often a parallel effort. The FPGA implementation of the design in such cases, helps identify functional bugs that might not have been possible in the normal ASIC design flow. This is made possible primarily because these FPGAs are tested in the actual environment intended for the ASIC. Also, finding a functional bug after fabrication of an ASIC is very expensive compared to finding the same in a FPGA prototype. All these factors have a significant impact on the design methodology for FPGAs.

Logic Synthesis tools for FPGAs differ from conventional ASIC synthesis tools. FPGA synthesis tools like *FPGA Compiler* from Synopsys have specific algorithms to exploit the architectural resources available on an FPGA. A few of the FPGA vendors currently supported by *FPGA Compiler* are Xilinx X4000, Actel ACT2, Altera Flex, and AT&T ORCA. The *FPGA Compiler* has specific optimization algorithms for these specified architectures. Further, Verilog or VHDL coding style can be tailored to the specific FPGA being used. Inefficient code used for synthesizing FPGAs can have an adverse impact on the quality of results in FPGAs.

## 7.2 Xilinx 4000 Architecture

Xilinx 4000 (XC4000) series FPGAs contain programmable logic cells known as Configurable Logic Blocks (CLBs) and Input/Output Blocks (IOBs). CLBs and IOBs are essentially Xilinx macros. The core of the Xilinx device is a matrix of identical CLBs. Each CLB contains programmable combinational logic and storage registers. The core also contains three state devices and decode logic. The combinational logic section of the CLB is capable of implementing any Boolean function of 5 or less variables and a subset of Boolean functions with 6-9 variables. The periphery of the Xilinx device is made up of IOBs which can be programmed as an input, an output (with or without three state control), or a bidirectional I/O cell. The *FPGA Compiler* directly maps HDL code to the Xilinx 4000 series CLB, IOB, and a limited number of special cells in the Xilinx 4000 library.

The XC4000 FPGAs are SRAM based. On power-up, the programmable FPGAs need to be configured by loading the configuration memory from an external memory. The XC4000 family has certain other dedicated resources on the chip like the JTAG IEEE 1149.1 Boundary scan logic, an industry wide standard for board level testing. You can choose to use this capability by instantiating JTAG components from the Xilinxlibrary. The XC4000 family has free routing resources available for high fanout nets which require minimum skew. These resources can be utilized by instantiating up to eight global clock buffers which drive eight global nets horizontally across the device.

XC4000 devices have dedicated resources for a Global Set/Reset net that can be utilized to initialize all the sequential elements inside the CLBs and IOBs on the device. This capability can be used, if your design has a global signal that effects every flop in your design, by instantiating a STARTUP component from the Xilinx library. However, to simulate the asynchronous behavior in the RTL code you must include the asynchronous capability in the code and later disconnect the asynchronous net and instead connect it to the STARTUP component.

Xilinx also has a library of XBLOX modules which are very efficient in terms of area and timing. Some of the XBLOX modules can directly be inferred by the *FPGA Compiler*. Others must be instantiated in your VHDL or Verilog code. For designs which contain Xilinx XBLOX modules, the Xilinx X-BLOX synthesis software needs to be run on those designs to generate a Xilinx optimized implementation for that XBLOX module. XBLOX synthesis from Xilinx uses Xilinx-specific optimization techniques and information about chip resources to produce fast and efficient circuit designs. Since Xilinx does not supply simulation models for the XBLOX modules, they are black boxes in the Synopsys environment. For simulation you will require a gate level netlist of the instantiated XBLOX modules. The XBLOX modules which can be currently inferred are ADD_SUB, COMPARE and INC_DEC. Some of the X-BLOX modules which can be instantiated are COUNTER, ACCUM, PROM, SRAM, ANDBUS, DECODE.

## 7.2.1 Configurable Logic Blocks (CLBs)

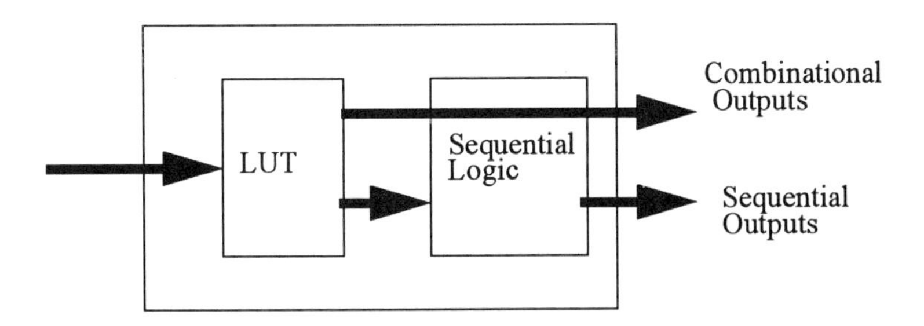


**Figure 7.1 CLB Architecture for XC4000**

The CLBs have four outputs: two registered and two combinational as shown in Figure 7-1. They consist of two 4-input function generators, F and G, and one 3-input function generator, H. These function generators are implemented as look-up tables. Each of the CLBs can be used to implement a RAM or logic. Combining these completely independent function generators, any 5-input or 9-input combinational function can be implemented in a CLB. Also, the CLBs contain two flip-flops with asynchronous set or reset and a clock enable signal.

### Latches and Flops

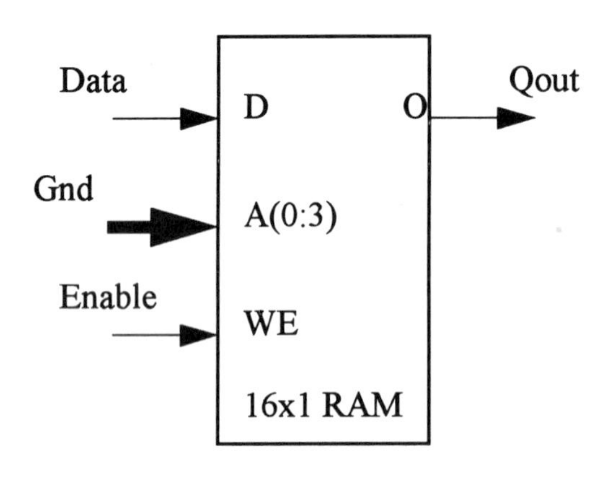


**Figure 7.2 16X1 RAM Configured as a Latch**

CLBs do not contain any latches. Hence, you must avoid inferring latches in your HDL code. Latches are inferred from HDL code when you do not specify the values of signals (VHDL) or registers(Verilog) under all possible conditions. If there are latches inferred in your code, they will be implemented using the function generators. This will result in combinational feedback loops in your designs. The recommended methodology to implement latches in Xilinx is to instantiate the 16x1 RAM primitive. Figure 7-2 shows the 16x1 RAM primitive configuration which implements the latch.

## Gated Clocks

Example 7.1 shows a VHDL and Verilog code template that can be used to infer flops with clock enables. The flops available in CLBs already have a clock enable pin, hence, one can use the code template in Example 7-1 for design scenarios that need gated clocks. The intent is to exploit all the dedicated logic resources available rather than use the limited combinational resources to implement the gated clock logic.

## Example 7.1 Inferring Flip-Flops with Clock Enable

### Verilog Code

```
always @(posedge clk)
begin
  if (ce) d = in1;
```

### VHDL Code

```
process
begin
  wait until clk = '1' and clk'event;
  if (ce = '1') then
      d <= in1;
  end if;
end process;
```

## Finite State Machines

The XC4000 has several flops, but narrow function generators. Hence, *one-hot* encoding schemes should be used for state machines. In one-hot encoding schemes, the number of flops equals the number of states in the design. Hence, there is no decoding logic required. In addition, the XC4000 architecture has dedicated wide I/O decoder resources on chip. These can be utilized by instantiating Xilinx edge decoder primitives in your HDL code.

## RAM/ROM

Xilinx XC4000 devices can efficiently implement RAMs and ROMs using CLB function generators. XC4000 libraries include 16x1 and 32x1 RAM and ROM primitives that you can instantiate in your code. These primitives must be instantiated in your HDL code to efficiently implement memory in the XC4000 family. In the

Synopsys environment, these primitives are black boxes and will remain as unresolved references. The Xilinx SYN2XNF program automatically merges the XNF file for the memory when translating the Synopsys generated XNF file to Xilinx XNF format as described in section 7.2.2.

### Clocks and Buffering

There are four vertical dedicated routing lines for each CLB column. Hence, there is an upper limit of four imposed on the number of clocks feeding a CLB column. Otherwise there are eight global horizontal dedicated routing resources available. The Synopsys insert_pads utility should automatically infer clock buffers for the clocks in your design. You may use these buffers for other fanout signals by either manually instantiating these pads in your HDL code or by inferring clock pad buffers for non-clock ports. This can be done by assigning the clock attribute to these ports using the set_pad_type -clock command.

### 7.2.2 Input Output Blocks (IOBs)

IOBs can be configured as either input, output, or bidirectional. When configured as an input, the input signals can be registered/latched or a direct input with pull-ups/pull-downs. When the IOBs are configured as an output, in addition to the above options, you can three-state the output and control the slew rate. The bidirectional IOBs have all the characteristics of input and output IOBs.

## 7.3 Synopsys Setup (.synopsys_dc.setup) For Xilinx

The following are the different variables that must be specified to synthesize a design using Xilinx technology libraries.

1. Include the path to your Xilinx libraries in your search path.

   search_path = search_path + <unix_path_to_xilinx_libraries>

2. Specify the Synopsys target_library, link_library, and symbol_library variables. The libraries required to target Xilinx 4000 FPGAs are as follows:

   xprim_4000-s.db

   Contains XC4000 primitive gates and flops whose delays do not vary with density of part. The "s" specifies the speed-grade.

   xprim_parttype-s.db

   Contains XC4000 cells like three-state buffers, clock buffers etc. whose delays vary with density of part.

   xgen_4000.db

   XC4000 cells that contain no timing information for example, pullups.

xfpga_family-s.db

Contains CLB and IOB primitives from the specified family.

xio_parttype-s.db

XC4000/A/H input/output buffers whose delays vary with the device type.

```
link_library = {xprim_4005-5.db xprim_4000-5.db xgen_4000.db
xfpga_4000-5.db xio_4000-5.db}
target_library = {xprim_4005-5.db xprim_4000-5.db xgen_4000.db
xfpga_4000-5.db xio_4000-5.db}
symbol_library = {xc4000.sdb}
```

3. Declare the XBLOX designware library (provided by Xilinx) in the `synthetic_library` variable. This allows *FPGA Compiler* to directly infer the supported XBLOX modules during synthesis.

```
synthetic_library = {xblox_4000.sldb}
```

4. Identify the location where the intermediate files for the XBLOX designware libraries are stored.

```
define_design_lib xblox_4000 -path <path_to_intermediate_files>
```

5. This variable affects the `compile` command in DC and *FPGA Compiler*. When set to `true` it prevents feed throughs and any net from driving more than one output port in the design. It does so by adding buffers.

```
compile_fix_multiple_port_nets = true
```

6. These variables control the naming of buses in the design.

```
bus_naming_style = "%s<%d>"
bus_dimension_separator_style = "><"
bus_inference_style = "%s<%d>"
```

7. This variable works with Xilinx XACT 5 version and above.

```
xlnx_hier_blknm = 1
```

8. This variable writes out version 5 xnf.

```
xnfout_library_version = "2.0.0"
```

## 7.4 Synopsys FPGA Compiler Flow

This section describes the complete design flow to be followed when synthesizing a VHDL/Verilog description to a Xilinx 4000 device using the *FPGA Compiler*. At each stage the relevant `dc_shell` commands are provided. The `.synopsys_dc.setup` described in section 7.3 should be used. Once you have coded your design in VHDL or Verilog and performed behavioral simulation of the design to verify its functionality, you are ready to read the design into Synopsys.

1. Read in your HDL code using the read command with the appropriate options. The read command performs syntax checking and basic code optimization such as, constant propagation and dead-code elimination before creating a network of logic equations and registers.

```
read -f vhdl <vhdl_file_name>
```

2. Specify timing and other optimization constraints on the design as shown in example dc_shell script below.

```
create_clock -period 30 CLK
set_input_delay 2 -clock CLK all_inputs() - CLK
set_output_delay 2 -clock CLK all_outputs()
set_load 1 all_outputs()
```

3. Set the port_is_pad attribute on the design to enable automatic pad synthesis. Pad synthesis is recommended before synthesis when targeting to Xilinx FPGAs because the IOBs contain sequential elements. During synthesis, the sequential elements from the core are absorbed into the IO pads if they are already present in the design. There are two basic advantages to this. First, this implies saving in the CLB resources. Second, this results in accurate routing delay estimates from the pad to the flop in the IOB.

```
set_port_is_pad current_design
insert_pads
```

4. Use *FPGA Compiler* to map the design to CLBs and IOBs. The compile command produces a netlist of CLB, IOB, TBUF cells, and a limited number of special cells (latches and adder/subtracter cells from the Xilinx DesignWare library) from the FPGA library.

```
compile
```

5. Write out a Synopsys .db file netlist of CLBs and IOBs.

```
write -hier -o pre_xnf.db
```

6. Run reports on the mapped design to find out the timing, total number of CLBs used, and the logic functions implemented in any specific CLB.

```
report_timing
report_cell
report_fpga
```

7. Convert the netlist of CLBs and IOBs to Xilinx primitive cells from the Xilinx primitive library. The Xilinx placement and route tools accept only a gate level description. To convert this netlist of CLBs and IOBs to a netlist of Xilinx

primitives, execute the replace_fpga command. The replace_fpga command enables one to view the actual logic within the CLBs and IOBs. There are options available with the replace_fpga command to control the hierarchy generated in the netlist. It is possible to create a separate level of hierarchy for each of the CLBs, IOBs, and the table lookups within the CLBs.

```
replace_fpga
```

8. To pass information about the Xilinx architecture being targeted to downstream Xilinx tools, specify the part attribute using the set_attribute command. Shown below is an example for 4005pc84.

```
set_attribute current_design "part" -type string "4005pc84"
```

9. You can associate each of the primitive cells in the netlist with a particular CLB in the XNF file by specifying the xnfout_use_blknames attribute on the design. If this information is missing in the XNF file then the Xilinx PPR tool is free to place the logic in any CLB. Setting this variable to false provides greater flexibility to the PPR tool though the binding chosen by the *FPGA Compiler* will be completely ignored.

```
set_attribute current_design "xnfout_use_blknames" -type boolean TRUE
```

Here is an example section from the XNF file which shows the BLKNM attribute.

```
SYM,BITS_SEEN_reg<1>,FDCE,HBLKNM=U125,SCHNM=FDCE,LIBVER=2.0.0
```

This above line in the XNF file binds the sequential element BITS_SEEN_reg<1> inferred from the HDL code to a CLB with instance name U125.

10. In the XNF file there exists a FMAP symbol for each of the F and G function generators for all the CLBs in the design. This gives information on the schematic implemented in each F and G function generator. If this information is missing in the XNF file, the Xilinx PPR program is free to move logic from F to G and/or from G to F within the same CLB.

```
set_attribute current_design "xnfout_write_map_symbols" -type boolean TRUE
```

Here is the relevant section from an XNF file that shows the mapping of a function generator in the CLB U125.

```
SYM,U178_map,FMAP,MAP=PUC,HBLKNM=U125,LIBVER=2.0.0
PIN,I1,I,sum204<1>
PIN,O,I,n321
PIN,I2,I,IS_LEGAL74
END
```

11. Write the design in the Xilinx Netlist Format (XNF) using the write command with the xnf option.

```
write -format xnf -hier -o post_xnf.sxnf
```

Use the Xilinx Partition, Place and Route (PPR) program to partition, place, and route the XNF netlist generated in step 11. This program produces a placed and routed LCA file that is used by the Xilinx MAKEBITS program to generate a bit stream for programming the XC4000 device. The placed and routed LCA file can also be translated into an XNF file using the lca2xnf program provided by Xilinx. This XNF file contains the post routed delay information.

The XNF (.sxnf) file from FPGA Compiler must be convered into a Xilinx XNF file using the syn2xnf program. Then, the Xilinx xmake program should be run on the xnf file generated from syn2xnf to get the .lca file. The xmake program consists of a series of Xilinx programs which can also be run serially in the event of any errors in the flow. The .lca file is the converted back to an .xnf file using the lca2xnf program. Finally this xnf file is converted to files which are required for VHDL simulation and incremental optimization. The xnf2vss program takes the .xnf file generated by lca2xnf and generates the following files: VHDL netlist (.vhd), SDF file (.sdf) and XNF netlist (.vxnf). The VHDL netlist and the SDF file can be used for gate level simulation. Further, the XNF netlist and the SDF file can be used for incremental optimization using DC.

### 7.4.1 FPGA Compiler to Xilinx Tools

User specified timing constraints and path exception commands in the Synopsys environment can be passed down to the Xilinx Partitioning Place and Route (PPR) tools through the Xilinx Netlist Format (XNF). This allows the PPR tool to have information about the critical paths in the design. The timing constraints can be specified in the Synopsys environment using the commands such as create_clock, set_input_delay, and set_output_delay as discussed in section 7.2.2. The path exception commands are set_false_path and set_multicycle_path. The Xilinx software supports four types of timing constraints: input port to output port (P2P), input port to register (P2S), register to register (C2S), and register to output port (C2P). Passing constraints to the XNF file increases the run time of the XNF writer and also that for the Xilinx PPR software. You can turn off passing constraints to the XNF file by setting the following variable to zero:

```
xnfout_constraints_per_endpoint = 0
```

## Recommended Further Readings

1. Field Programmable Gate Arrays. Kluwer Academic Publishers. *Stephen. D. Brown, Robert J. Francis, Jonathan Rose, Zvonko G. Vranesic.*

2. Synopsys FPGA Compiler Reference Manual

# CHAPTER 8 *Design for Testability*

The ever-increasing density of ASICs, the whole-sale switch to surface-mount technology, and the growing interest in multi-chip modules (MCM), have resulted in testable designs becoming a greater priority. Thus far, designers have considered testability as an issue which comes into play at the very end of the design cycle. However, in the ASIC design flow based on synthesis, it is essential that designers develop a test strategy and address testability issues concurrently with other activities in the design cycle. In this chapter, Test Synthesis and Automatic Test Pattern Generation (ATPG) using the Synopsys *Test Compiler* (TC) are discussed.

The chapter begins with a brief introduction to test synthesis. This is followed by a description of the specific steps involved in test synthesis using TC. Then, several design specific issues which impact the approach to test synthesis are described, followed by tips for users. Finally, discussions on several "classic scenarios" faced by designers when using TC are provided.

## 8.1 Introduction to Test Synthesis

Logic synthesis results in a netlist which usually contains sequential non-scan cells and other combinational gates from the technology library. The primary objective of test synthesis is to improve the observability and controllability of the design. In other words, one must be able to detect manufacturing defects in the design. In general, running functional vectors on the tester does not give more than 50-60% fault coverage. Hence, one approach is to run ATPG after fault simulation to target the remaing faults.

The commonly used fault model in the industry is the stuck-at-fault model. Any manufacturing defect in this model is represented as either a stuck-at-0 or stuck-at-1 fault. This helps make the testing process completely technology independent with every node/pin in the design having two faults associated with it (stuck-at-0 and stuck-at-1.)

*Scan* technique is the most widely used Design For Testability(DFT) technique, and more importantly, is supported by most test synthesis tools. This technique involves replacing the sequential non-scan cells by scan cells of the desired scan style . This transformation enables the sequential scan cells to be connected as a shift register in the scan mode. Further, for ATPG each of these scan cells behave as a pseudo primary input as well as a pseudo primary output. In this section, we discuss four basic issues related to Test Synthesis using TC, namely, Scan methodology, Scan Style, Scan Insertion and ASIC Vendor issues.

### 8.1.1 Scan Methodology

*Full Scan* and *Partial Scan* are the two design methodologies supported by most test synthesis tools. In a full scan methodology, all the sequential cells in the netlist are replaced by scan cells. On the other hand, in a partial scan methodology, only some of the sequential cells are replaced by scan cells. The choice of the sequential cells to be replaced by scan equivalents, is made based on area and timing constraints required by the design. The user can also specify if fault coverage, timing, or area is of highest priority.

Full Scan designs in general, achieve a higher fault coverage when compared to partial scan designs. The fault coverage of a partial scan design is dependent on the number of non-scan sequential cells replaced by scan cells. Further, design specific characteristics like sequential depth and sequential feedback loops, impact the fault coverage achieved using partial scan.

### 8.1.2 Scan Styles

In TC terminology, the scan style refers to the kind of scan cell used. The four commonly used scan styles are as follows:

1. Multiplexed flip-flop
2. Level sensitive scan design
3. Clocked scan cell
4. Auxiliary LSSD cell

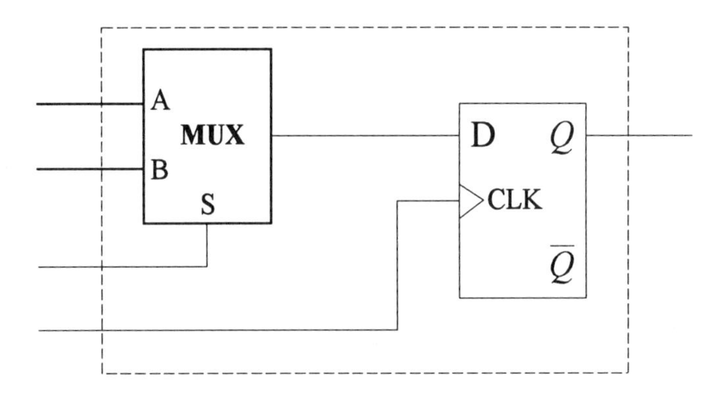


**Figure 8.1 Muxed Scan Flop**

Figure 8.1 shows a multiplexed flip-flop scan cell. In this chapter, we discuss only the *multiplexed flip-flop scan style*. However, most of the test design rules discussed are also applicable to other scan styles. This scan style is supported by most ASIC vendors. For a multiplexed flip-flop scan style the scan ports required are the *scan-in*, *scan-enable*, and *scan-out* ports. The normal clock is used in the test mode in this scan style.

## 8.1.3 Scan Insertion

Scan cells have two different modes of operation: *normal* and *scan*. In normal mode, the scan cell's functionality is same as that of the sequential non-scan cell. In scan mode, the scan cells are linked in the form of a shift register.

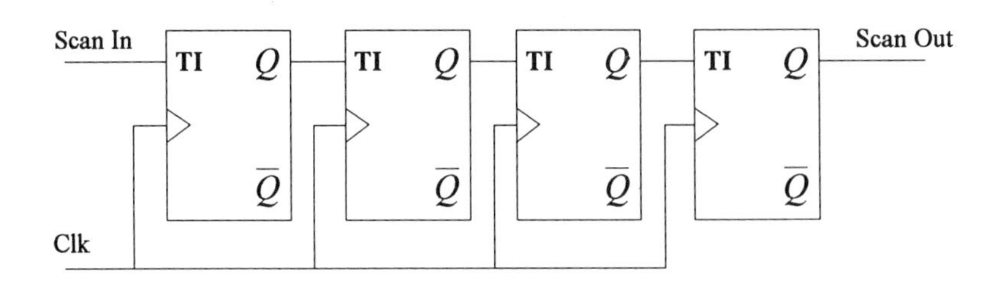


**Figure 8.2 Scan Cells Linked to form a Scan Chain**

When scan cells are linked to form a *scan chain* as shown in Figure 8.1, all the scan cells are controllable and observable. Since shifting of data into the scan chain is performed serially, it takes N clock cycles to shift an entire pattern into the scan chain, where N is the maximum length of the scan chain. Configurations with multiple scan chains are supported by most synthesis tools.

Scan insertion results in design overheads such as, the use of extra scan ports, an increase in silicon area due to use of scan flops, and greater timing delays due to the insertion of the scan cells for the sequential non-scan cells. It is possible to reduce the port overheads by sharing the scan ports with functional ports.

After inserting the scan logic in the design, the ATPG algorithm is used to generate test patterns. Full scan ATPG algorithm is combinational, while the partial scan ATPG algorithm is sequential. Test patterns can be generated in the format supported by the simulator used to simulate the test vectors.

### 8.1.4 ASIC Vendor Issues

The test strategy used for a design is almost completely dependent on the requirements of the ASIC vendor. In other words, the requirements of the ASIC vendor must be taken into consideration prior to deciding on the test synthesis strategy. Listed below are some of the critical issues which involve the ASIC vendor.

1. What scan style does the ASIC vendor support?

   ASIC vendors usually support only some scan styles and not all the available scan styles. Most ASIC vendors support the multiplexed scan flip-flop style.

2. How many clocks are supported by the tester when in test mode?
3. Is there a limit on the number of waveforms supported by the tester?
4. Is there a limit on the number of scan chains allowed?

   Most ASIC vendors impose a restriction on the number of scan chains.

5. Is there a limit on the length of the scan chain?
6. Is sharing of functional ports with `test_scan_in` and `test_scan_out` ports supported?
7. Does the vendor require that during scan-shift all the outputs have no switching, all outputs are three-state outputs.
8. Does the vendor library support automatic pad synthesis?
9. What is the format of the test vectors required by the vendor?
10. Does the vendor accept parallel vectors or serial vectors for sign-off simulation? (Applicable only to VHDL and Verilog formats.)

11. What is the maximum number of scan bits supported by the vendor's tester?

    The total number of scan bits is simply the number of scan vectors multiplied by the number of flip-flops in the scan chain.

12. Do the formatted vector files require a specific naming convention?
13. Is there a limit on the size of the vector files?

# 8.2 Test Synthesis Using Test Compiler

In this section, we first provide a brief introduction of the commonly used TC commands. The purpose of this section is to familiarize the reader with test synthesis using TC. Then, the methodology for test synthesis using TC is presented. All test synthesis concepts discussed in the remaining part of this chapter are specific to TC.

## 8.2.1 Commonly Used TC commands

The definitions of the basic TC commands provided in this section, should help the reader understand the TC flow with regard to actual dc_shell commands.

### check_test

This command infers a default test protocol and performs a DRC check by simulating the test protocol. One must execute the check_test command before scan insertion as well as after scan insertion. check_test is the main debugging command in TC and it's output must be thoroughly analyzed for its implications on the fault coverage.

### create_test_clock

This command is similar to the create_clock command for DC. TC automatically infers clocks during check_test by backtracking from the clock pins of registers. The create_test_clock command is used to specify the waveform and clock period in the test mode.

### insert_scan

The insert_scan command replaces the non-scan sequential cells with scan equivalent cells and connects the scan cells to form a scan chain.

### create_test_patterns

This command is used to generate the test patterns for the specified design. The command also writes out a binar vector database (.vdb file) in the current working directory.

### set_test_hold

This command is used to specify static values on primary input ports throughout test mode.

### set_test_dont_fault

This command is used to remove specific faults from the master fault list. One application could be a RAM, since in general, ASIC vendors provide vectors for RAMs.

## 8.2.2 Identifying Scan Ports

TC uses the signal_type attribute to identify scan ports. Functional ports can be identified as scan ports, by assigning this attribute using the set_signal_type command. TC creates scan ports automatically if no functional ports are identified with the signal_type attribute. In the muxed flip-flop scan style, where normal clock is used as test clock, one must not associate a signal_type attribute, *test_clock* with the clock port.

## 8.2.3 Test Synthesis Flow Using Test Compiler

In this section, the steps involved in test synthesis using TC are outlined.

1. Read in the HDL code of the entire design into DC. In the example dc_shell command shown below, VHDL_FILES is a variable which is assigned to a number of VHDL files.

   ```
   VHDL_FILES = {A.vhd B.vhd C.vhd TOP_LEVEL.vhd}
   read -f vhdl VHDL_FILES
   ```

2. Set your current_design to the top level and specify the test methodology and scan-style.

   ```
   current_design TOP_LEVEL
   set_test_methodology full_scan
   set_scan_style multiplexed_flip_flop
   ```

During the synthesis phase, ensure that DC does not use any sequential cells which do not have scan-equivalents of the desired scan style in the technology library. There are two ways to ensure this. One approach is to place a dont_use attribute on all the scan cells in the library as well as all non-scan sequential cells which do not have a scan equivalent.

Another simpler alternative is to use the *Test Smart Compile* approach. In this approach the user must specify the scan style before compile. The Test Smart Compile is turned on by specifying the scan style (using set_scan_style command) before compile. Then, DC automatically ensures that only those sequential cells which have scannable equivalents in the specified scan style, are inferred during optimization. However, for partial scan designs, users may choose to use non-scan cells functionally in critical paths, and choose not to replace them during scan insertion.

3. Synthesize your design after specifying area and timing constraints. One can use the compile strategies discussed in Chapter 4.

```
include constraints.scr
compile
```

4. Specify the timing related test attributes as shown below. For the default values refer to the TC Reference Manual. It is recommended that these variables be specified in your .synopsys_dc.setup file to ensure consistency throughout the design project.

```
test_default_period = 1000.0
test_default_delay = 50.0
test_default_bidir_delay = 550.0
test_default_strobe = 950.0
create_test_clock clock_port_list -waveform {450.0 550.0}
```

5. Specify the test methodology, scan style, number of scan chains, mixing of clock domains in the scan chain etc. using the set_scan_configuration command.

6. Analyze the testability of the design prior to scan insertion using the check_test command. A default test protocol is inferred and simulated on executing the check_test command. The default test protocol consists of the following: initialization vectors, scan-in/scan-out, parallel measure, capture and scan out strobe.

```
check_test
```

7. Save the design database in db format prior to inserting scan.

```
write -f db -hier current_design -output top.db
```

8. Perform Scan Insertion using the insert_scan command.

```
insert_scan
```

9. The next phase involves testability analysis after scan insertion. Analyze the testability of your design using the check_test command.

   ```
   check_test
   ```

10. Execute ATPG on a sample fault list to check for any ATPG conflicts which might exist. Refer to the section on ATPG Conflicts to understand and debug this situation. The command shown below generates test patterns for 5% of the faults in the design.

    ```
    create_test_patterns -sample 5
    ```

11. The next step is the JTAG synthesis phase. Group all the core logic except, three-state cells associated with three-state and bi-directional ports, into a separate level of hierarchy.

    ```
    group -design_name Core -cell_name top filter(find(cell,"*") -except
    {three_state_cell_list})
    ```

    The variable three_state_cell_list, is a user-defined variable which lists the instances of three-state cells.

12. Set current_design to the top level of the design hierarchy.

    ```
    current_design TOP_LEVEL
    ```

13. Specify the order of the boundary scan register (BSR) cells using the set_jtag_port_routing_order command.

    ```
    set_jtag_port_routing_order {list_of_ports}
    ```

14. Perform JTAG insertion with the required options. Use the -no_pads option if the ASIC vendor library does not support automatic pad synthesis. In general, to have complete control over the pad cells, it is recommended that you instantiate pad cells in your design.

    ```
    insert_jtag -no_pads
    ```

15. Perform testability analysis after JTAG insertion using the check_test command. The synthesized JTAG logic has quite a few testability violations associated with it; such as uncontrollable clocks. Hence, you may either ignore the warnings associated with the JTAG logic or specify a test_isolate attribute on the sequential cells in the JTAG logic to prevent these warning messages.

    ```
    check_test
    ```

16. Save the db file after JTAG synthesis.

    ```
    write -f db -hier current_design -output after_jtag.db
    ```

17. Group all the JTAG logic into a separate level of hierarchy, and assign a test_dont_fault attribute on them, to avoid being considered in the fault coverage calculation. The design should now consist of the core instance surrounded by all the JTAG logic and three-state buffers. Assuming the instance name of the core design is CORE1, the dc_shell script shown below describes the steps.

```
/* Finding all cells at the Top Level */
find(cell, "*")
cell_list = dc_shell_status - CORE1
/* Finding all cell instances besides three-state cells */
filter cell_list "(@is_combinational==true)||(@is_sequential ==
true)||(@is_hierarchical == true)||(@is_black_box == true))"
/* Group all the JTAG logic */
group -design JTAG -cell JTAG1 dc_shell_status
/* Putting a test_dont_fault attribute on the JTAG logic */
set_test_dont_fault JTAG1
```

18. In order to control and specify the characteristics of the desired pad cell, use the set_pad_type command. DC inserts pads for all ports in the design which have the port_is_pad attribute. This attribute can be applied using the set_port_is_pad command.

```
set_port_is_pad current_design
set_pad_type
insert_pads
```

19. Execute ATPG using the following command. This creates a .vdb file which is a binary vector file. TC does not generate test patterns for the JTAG logic, hence, additional functional vectors will be required for the JTAG logic.

```
create_test_patterns
```

20. Finally, generate the test vectors in the required format.

```
write_test -format <format supported by TC>
```

## 8.3 Design-Specific Issues in Test Synthesis

The methodology discussed in section 8.2 should give the reader a sound understanding of the steps involved in test synthesis using TC. However, testability issues are often design-specific and might require special strategies. Thus, when building testability into the design, one must consider several issues such as reset signals, illegal paths, designs with RAMs, designs with latches, three-states, scan chains, and scan chain ordering. In this section, these issues have been discussed with a special focus on how they can be tackled using TC. A prior exposure to design for testability should help the reader better understand the tool related solutions discussed.

### 8.3.1 Testability Analysis at the Module Level

Testability analysis works best with a divide and conquer approach. In other words, always analyze the testability of the design at the module level rather than directly at the top level of a hierarchical design. This is similar to *compile* in DC. Just as one would *compile* smaller blocks of logic and integrate them using a bottom up compile strategy, when using TC, it is recommended that one analyze smaller blocks of logic for testability.

Testability analysis at the module level helps isolate testability issues at the early stage of the design cycle. This helps simplify the debug process. At the module level, it is also recommended that one run ATPG on a small sample fault list to detect any ATPG conflict situations using the command shown below:

```
create_test_patterns -sample 3.
```

### 8.3.2 Clocks and Reset Signals

In the testmode, it is important that all clocks and asynchronous reset signals be controllable from the primary inputs. For internally generated clocks in the design (clock dividers, for instance), one approach is to synthesize a mux in the HDL code such that, in the testmode, the test clock from the primary input is selected. The same approach can be followed for the reset line.

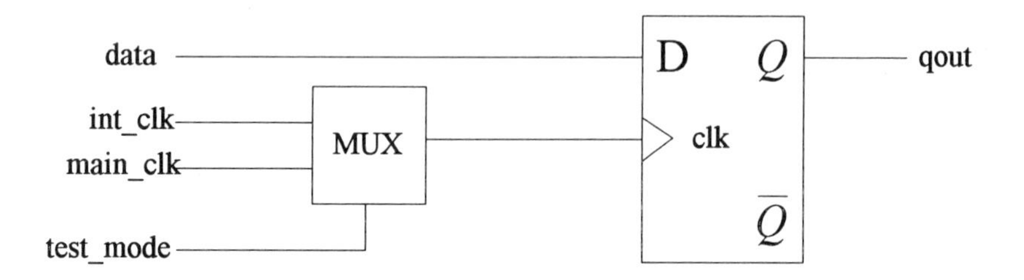


**Figure 8.3 Flip-flop Driven by Muxed Clock Signals**

The flip-flop with a muxed clock signal shown in Figure 8.2 can be inferred using the VHDL/Verilog code shown in Figure 8.1. In this code segment, when the signal `testmode` is high, the controllable clock from the primary input is selected, while in the functional mode the normal (internally generated) clock is selected.

### Example 8.1

**VHDL Code**

```
process(testmode, int_clk, main_clk)
begin
if (testmode = '1') then
    muxed_clk <= main_clk;
else
    muxed_clk <= int_clk;
end if;
```

**Verilog Code**

```
assign muxed_clk = (testmode == 1'b1) ? main_clk : int_clk;  // mux
```

## 8.3.3 Designs With Latches

For designs with latches, one must consider the different cells available in the technology library for the scan style selected. In other words, one must ensure that the scan style chosen, has scannable equivalents for latches in the technology library. LSSD is one scan style which has scan equivalent for latches. Latches can also be treated as transparent in the testmode. TC uses a mux model for latches in the transparent mode. The latch is replaced by a 2-to-1 mux model with the enable line acting as the select, and the two inputs being data signal and the X signal. Thus, when the latch is enabled, the data line is selected, and when disabled, the X is propagated. The TC ATPG automatically determines the values required to make the latches transparent during test. In the transparent mode, there is a potential for combinational feedback loops, if there exists a path from the output of the latch to the input of the latch. Latches can be placed in the transparent mode using the following command.

```
set_scan false -transparent cell
```

### 8.3.4 Combinational Feedback Loops

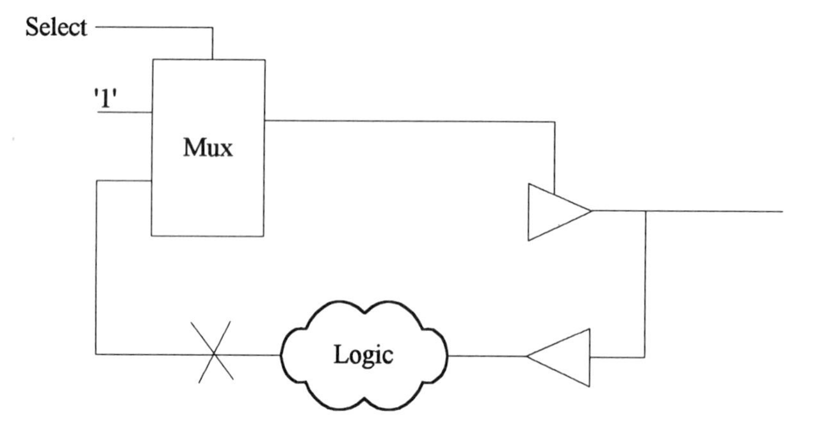


**Figure 8.4 Combinational Feedback Loop**

TC identifies and breaks combinational feedback loops automatically. The selection of the point in the combinational feedback loop chosen is reported in the check_test output. Heuristics have been incorporated into check_test that determine the point at which combinational feedback loops are broken. Further, TC understands functionally broken loops. The nodes at which check_test avoids breaking combinational feedback loops in order of priority are as follows: tristate/bidir enable pins, pins directly feeding asynchronous inputs of sequential elements, pins directly feeding other inputs of sequential elements, and gates with a large fan-out.

### 8.3.5 Three-State Logic

By default, the insert_scan command adds disabling logic to three-state buffers. This ensures that during scan shift, only one driver is active on the three-state net. You can turn off the addition of this disabling logic using the -no_disable option of insert_scan ATPG does not check for contention/float during scan-shift and only during the parallel cycles. Hence, it is highly recommended that you let TC synthesize this disabling logic. However, if you can ensure that there shall be no contention or float during scan shift in the source VHDL/Verilog code, you may turn off this feature.

Bi-directionals are also treated as three-state nets by TC. Hence, the insertion of disabling logic would occur for bi-directionals unless the -no_disable option is used. Bidirectional ports, by default, are always configured in the output mode during test. There are options available with the set_scan configuration command to specify the direction of the bi-directional ports during scan shift.

### 8.3.6 Designs With RAMs

Designs with embedded RAMS often require special strategies for test synthesis. If the design contains embedded RAMs then one must consider one of the following test strategies. RAMs are generally black boxes in Synopsys environment because they cannot be functionally described in the Synopsys library. Starting v3.4a there have been enhancements made to Synopsys Library Compiler which allow such complex cells to be modelled. The new feature is due to a new capability called UNIGEN. Most ASIC vendors provide test vectors to test RAMs. Hence if the RAMs do not have a functional model then the objective during ATPG is to limit its X propagation in the rest of the design. Some of the common approaches to handle RAMs are as follows:

- Bypass the RAM outputs in the testmode using muxes at the outputs so that in the testmode the mux propagates a constant value.
- Use a register around the RAM (register bounding technique) to control the propagation of Xs (unknowns) in the design. Using a register ensures the observability of the inputs of the RAM and controllability of the outputs of the RAM.
- If the RAM has a feedthrough mode of operation, create a feedthrough library model of a RAM. Then, using the initialization test protocol, configure the RAM in this feedthrough mode.

An initialization test protocol is used to configure the ASIC for the test mode by providing certain initialization vectors. Note that an initialization test protocol can only be used when the RAM model is not a black box in Synopsys. Toggling the inputs of a black box will not result in any improvements in fault coverage. If the RAM model in the Synopsys library has setup/hold timing arcs between the data/address pins and the write/read enable and RAM select pins, one must place a test_isolate attribute on it. Without a set_test_isolate on the RAM, check_test will try to trace back the write/read enable and RAM select pins to infer a clock which may cause other DRC violations.

The set_test_assume command can be used to get an estimate of fault coverage. This allows the user to perform analysis of the improvements in fault coverage by initializing the RAM to different states. TC does not verify these values and the user must ensure that the RAM is initialized to that state. In addition, for each of the above approaches, a test_dont_fault attribute should be placed on the RAM cell. This prevents the inclusion of faults associated with the RAM cell in the fault coverage calculation.

### 8.3.7 Timing Attributes

The timing attributes for scan testing of the design should be appropriately set. The variable test_default_bidir_delay should be set to a value greater than the rising edge of the capture clock. This variable specifies the time when data is applied/released at the bi-directionals when in the input mode. If this variable is set before the active edge of the clock, there could be simulation mismatches during test vector simulation.

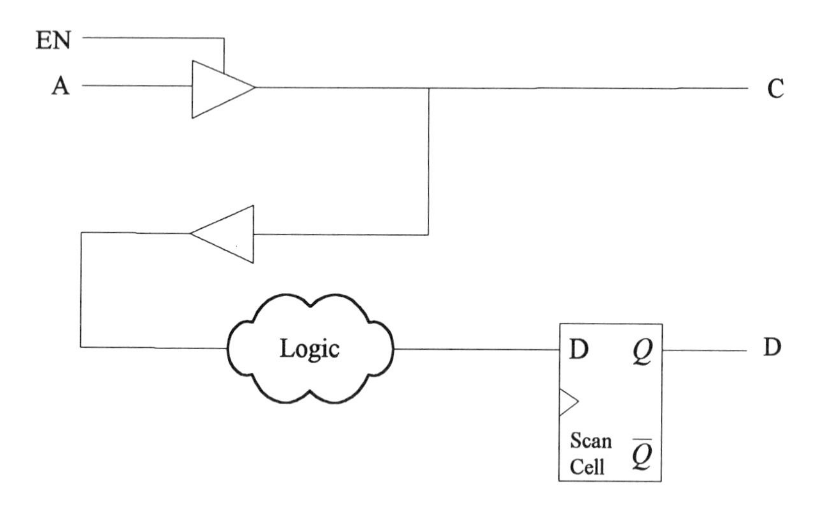


**Figure 8.5 Design with Bi-directional Paths**

Figure 8.5 shows a design with a functional path from the bidirectional port C to registers.

The default test protocol inferred by TC has four phases: Scan Shift, Parallel Measure, Parallel Capture, and Scan-Out Strobe. Each test pattern has all these phases. The length of the scan shift phase is equal to the length of the longest scan chain in the design. The remaining three phases are of one cycle with the test clock being pulsed in only during scan shift and the parallel capture cycle. Shown below are the values of the bi-directionals in the different phases:

Scan-Shift : M (Masked)

Parallel Measure : Pio (Input or Output)

Parallel Capture : M (Masked)

Scan-out Strobe : M (Masked)

If the bi-directional is used as an input in a particular pattern the tester applies data to the bidirectional data ports in the Parallel Measure cycle at the time specified by the variable test_default_bidir_delay. In the capture cycle the tester must release the data applied at the time specified by test_default_bidir_delay. If the test_default_bidir_delay occurs before the active clock edge, the bidirectional will float in the capture cycle before the data has been captured into the scan cells, resulting in an incorrect value being captured. This is the most common cause of simulation mismatches when simulating the ATPG generated test patterns. Hence, the restriction that test_default_bidir_delay must be set after the active clock edge.

If there is an ASIC vendor restriction that test_default_bidir_delay occur before the active clock edge, you have the following options:

1. Configure all the bidirectionals in the output mode throughout the testing phase. This can be achieved by setting the atpg_bidirect_output_only variable to true before executing create_test_patterns. This will however, result in lowering the fault coverage.
2. Use a strobe before the clock protocol. This protocol is automatically inferred by TC when the test_default_strobe is set before the active clock edge. In this protocol file the parallel measure and the parallel capture cycles are combined into one.
3. Use the TC variable, test_no_three_state_conflicts_after_capture to hold the values of the bi-directional ports in the parallel capture cycle from the parallel measure cycle.

### 8.3.8 Scan-In and Scan-Out Ports

Scan-in and scan-out ports should only be shared with functional ports if supported by the ASIC vendor. In general, sharing functional ports for scan-in ports causes extra loading of the ports, and sharing scan-out ports results in extra delay equivalent to a delay through a mux. When sharing ports, it is recommended that functional ports which do not have very tight timing requirements be selected.

## 8.4 Clock Skew

TC has no information of the timing in technology library cells. Hence, if there is any skew between the clocks, this information should be provided in the clock waveforms using the create_test_clock command. This is not possible if all the clocks are muxed into a single test clock in the test mode. This could be because of vendor tester clock support limitations, in which case, TC will not have the required skew information. There are two potential problems due to clock skew which can be detected only during post-scan timing analysis, or when simulating vectors - the problem of diverging scan chains diverging scan chains and illegal paths.

### 8.4.1 Diverging Scan Chains

Clock skew in the scan path could result in a scan pull-through or a *diverging scan chain*. In other words, during the scan shift phase, due to clock skew the same scan bit can be loaded into two successive scan cells.

By default, TC creates a separate scan chain for every clock domain in the design. The rising and falling edges of the same clock are treated as two different clock domains. If you need to mix clock edges on the same scan chain, it is recommended that for a *Return-To-Zero* clock you ensure all negative edge triggered flops occur before the positive-edge triggered flops in the scan chain. Similarly, for a *Return-To-One* clock all positive edge triggered flip-flops must occur before negative edge-triggered flip-flops. However, an issue could arise when the clock source is the same but different clock branches have different delays. Since TC has no timing information, check_test will not identify such timing problems. Post-scan timing analysis and/or simulating your scan-check test vectors will identify such problems.

The set_scan_path command can be used to specify which scan chain the flops in a sub-design must be assigned to, depending on the clock skew on that branch. Further, this command can also be used to explicitly order scan cells within a scan chain. This command provides a means to arrange the scan-cells in the scan-chain in the order of reversed skew.

### 8.4.2 Retiming Latches

Retiming latches can be used in the scan chain where there are hold problems identified on the scan path (Q -> Si), due to mixing of clocks or clock skew. One configuration of *retiming latches* is supported by TC. TC by default, automatically adds retiming latches when mixing clock domains on a scan chain or when manually specified by the user.

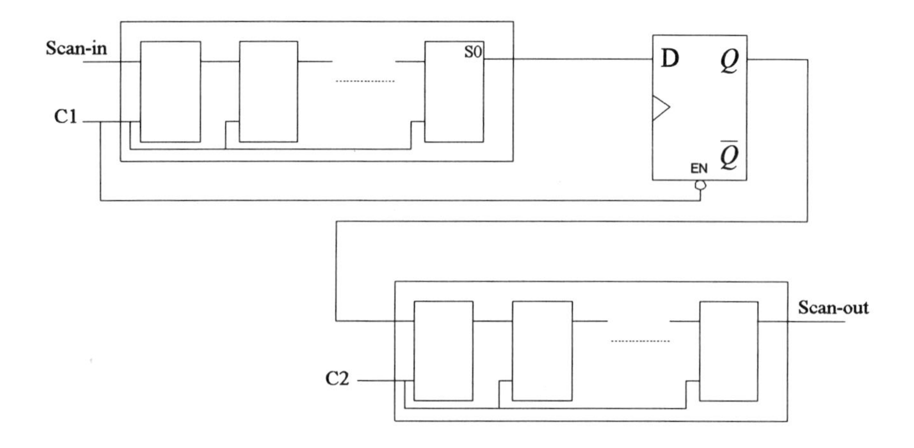


**Figure 8.6 Retiming Latch**

Consider a design with a scan chain and two clock trees as shown in Figure 8.6. Notice that the upstream flip-flops are driven by clock branch C1, and the downstream flip-flops by clock branch C2. When going from clock domain C1 to clock domain C2, a retiming latch is inserted between the *scan-out* pin of the last flop in the scan chain clocked by C1, and the *scan-in* pin of the first flop in the scan-chain clocked by C2. The enable of the latch is connected to C1 such that it is transparent when C1 is low. The latch holds the previous scanned value for the duration when the clock pulse is high. This approach works, provided the skew is not greater than the high pulse width of the clock. Alternatively, one must use DC to insert delays to fix the hold time violations.

### 8.4.3 Illegal Paths

Designs with datapaths from one clock domain to another have a potential for *illegal paths*. TC identifies an illegal path when the data being captured depends on the newly captured value rather than the scanned-in value. If the illegal paths are due to clock skews, TC does not detect this as a problem. Such situations can be detected through timing analysis or by simulating the test vectors. Under all circumstances, illegal path violations imply design problems and most often require re-design. Re-ordering the scan cells will not prevent illegal path violations on the datapath.

Illegal path violations can cause a substantial decrease in fault coverage, since the violating scan-cells will not capture data, that is, they are scan controllable only. One approach to avoid this, is to place the clocks in different capture clock groups, such that they are not clocked in the same cycle in the capture mode. Clocks are placed in the same capture group if they can be clocked in the same parallel capture cycle such that the values captured in the scan cells are consistent with those predicted by ATPG. TC will automatically place clocks in different capture clock groups if required, provided there are different test clocks specified in the test mode. However, there is an exception to this automatic approach. When there are muxed clocks in the testmode, there are likely to be skews between the different clock branches and TC cannot place them in different capture clock groups. In such cases, one must identify hold violations by performing timing analysis on the design. Delays can be inserted to fix hold violations by using the fix_hold DC command. For designs with multiple clock domains, the clocks might be asynchronous in the functional mode. But in the test mode, since all the clocks are active in every cycle, when performing timing analysis, one must check for hold violations on all paths.

In designs where there are multiple clock domains and datapaths between the clock domains, TC will place clocks in separate capture clock groups, provided they are independent clocks sources in the testmode and not multiplexed to one test clock. Further, the clock skew must be accounted for in the clock waveforms.

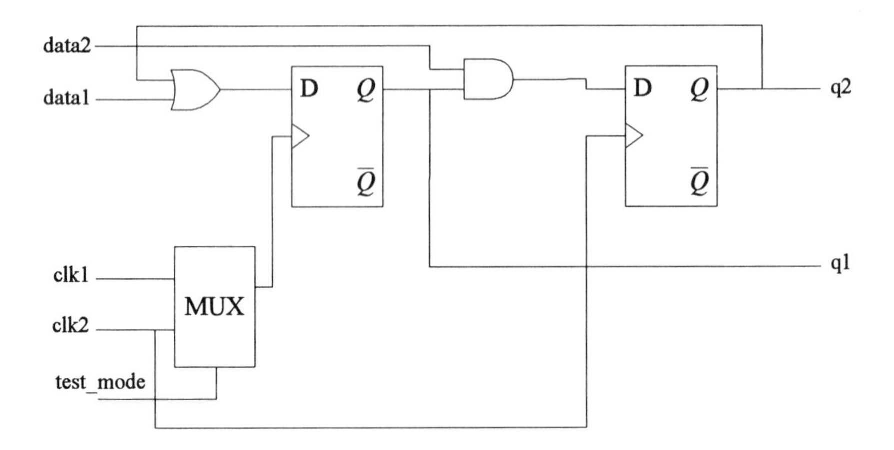


**Figure 8.7 Design showing Illegal Paths**

Figure 8.7 demonstrate the issues related to capture clock groups and illegal paths. This design consists of two flip-flops. Depending on the value of the testmode port during test, the two flops will be clocked by either the same clock clk2, or by clocks clk1 and clk2. Also, there are datapaths between clock domains clk1 and clk2 as well as between clk2 and clk1. We consider the following three possible scenarios that could arise:

**Scenario 1**

When the test_hold attribute of 1 is on the testmode port.

In this case, TC infers only one clock clk2, and will not account for any skew on the different branches of clk2. You must perform timing analysis of the design and identify the illegal paths in the design. Shown below is the relevant portion of the check_test output showing the inferred capture clock groups.

```
...creating capture clock groups...
Information: Inferred capture clock group : clk2. (TEST-262)
...binding scan-out state...
...simulating serial scan-out...
```

**Scenario 2**

When the test_hold attribute is 0 on the testmode port.

This will imply two clocks, clk1 and clk2, in the testmode. If the waveforms of the two clocks are identical, TC will place both clocks in the same capture clock group. Further, TC will clock both at the same time in the same capture cycle.

```
...creating capture clock groups...
Information: Inferred capture clock group : clk1, clk2. (TEST-262)
...binding scan-out state...
...simulating serial scan-out...
```

Clocks clk1 and clk2 are activated in the same capture cycle because TC does not have information on the timing skew between the two clocks. In this case, although the check_test results do not indicate any specific problems, there could be problems in the capture cycle due to skew in the different clock branches.

**Scenario 3**

When testmode has a test_hold attribute of 0 and has specified skewed clock waveforms.

In this case, TC infers two clocks, clk1 and clk2, in the testmode. If the test clock waveforms are specified taking into account the skew (after checking the skew of the clocks with the ASIC vendor), TC will be able to identify the illegal paths on the datapaths and place the clocks in different capture clock groups. Hence, in Figure 8.2, if we change the waveform of the clock clk1 from the default as follows:

```
create_test_clock clk1 -period 100 -wave {40 60}
```

the following capture clock groups can be observed.

```
 ...creating capture clock groups...
Information: Inferred capture clock group : clk1. (TEST-262)
Information: Inferred capture clock group : clk2. (TEST-262)
  ...binding scan-out state...
  ...simulating serial scan-out...
```

Also, in the test patterns generated, only one of the two clocks, either clk1 or clk2 is clocked in the capture cycle of each test pattern. Because, in this design there are datapaths between clock domains clk1 and clk2, as well as between clk2 and clk1.

TC does not consider clock skew when determining clock capture groups. However, the TC variable test_capture_clock_skew has been provided to specify a coarse estimate of clock skew. This variable can take on one of the following three values: no_skew, small_skew, and large_skew. The behavior of no_skew is identical to that discussed above. When set as small_skew interacting clocks with identical waveforms are put in different capture clock groups. large_skew is very pessimistic and all clock sources will be placed in different capture clock groups resulting in larger number of test patterns.

### 8.4.4 Scan Chains

By default, TC allocates all scan cells clocked by the same clock, to the same scan chain. Also, scan cells clocked by different edges of the clock are placed in different scan chains. Hence, if one has only one test clock in the testmode, because of multiplexing the functional clocks, TC will place all the scan cells in the same scan chain, by default.

### 8.4.5 ATPG Conflicts

When TC cannot generate any test patterns without avoiding contention/float or when unable to satisfy conflicting test assertion requirements, TC reports 0% fault coverage along with the ATPG conflicts. Currently the report_test -atpg_conflicts command cannot distinguish between contention and float. Hence, use the options with create_test_patterns to turn off contention and float independently to find out the cause of ATPG conflicts. Shown in Figure 8.8 are some of the common scenarios which causes ATPG conflicts.

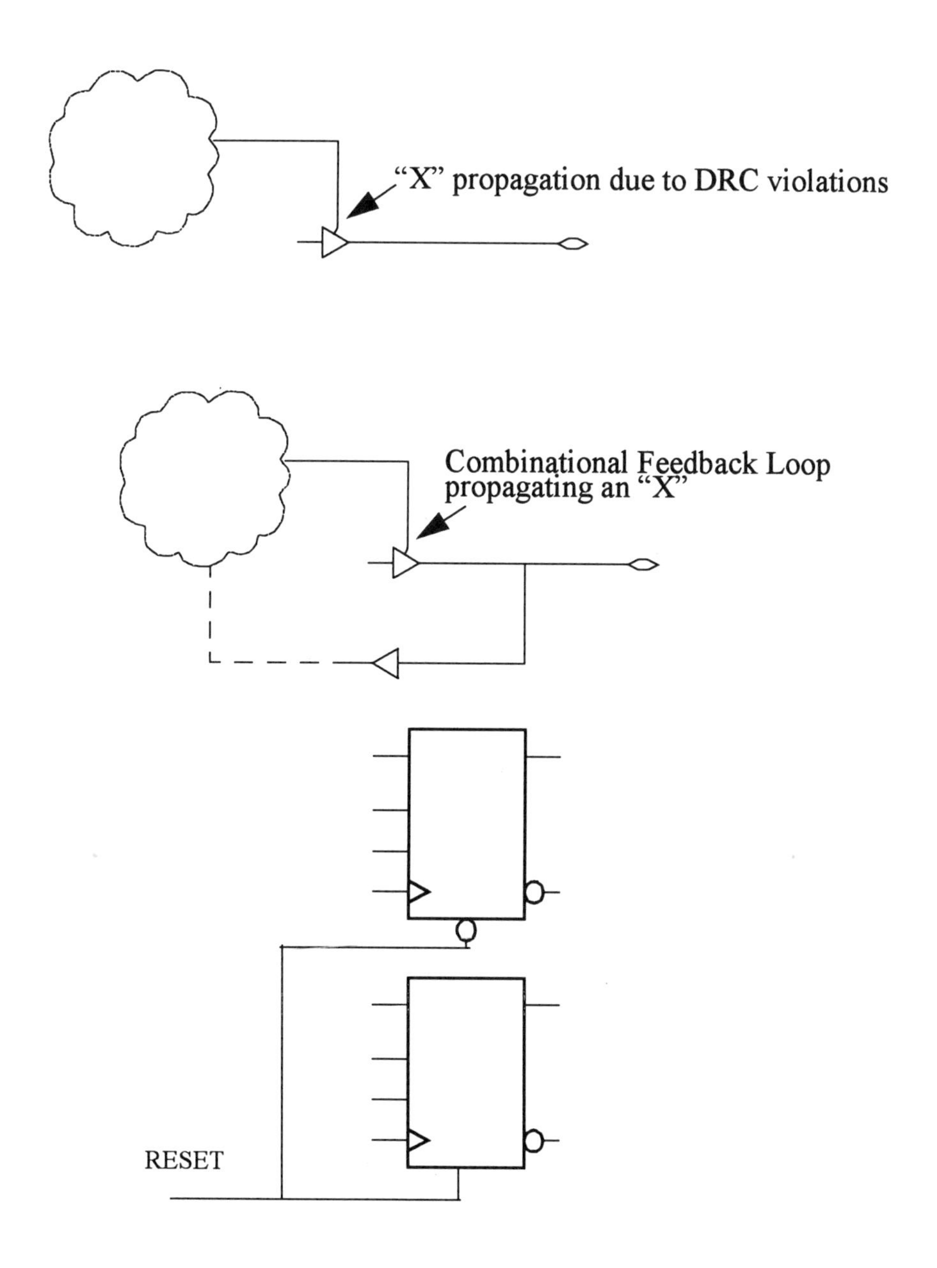


**Figure 8.8 ATPG Conflict Scenarios**

## 8.5 Test Compiler Default Test Protocol

Every test pattern consists of the following phases: scan-shift, parallel measure, parallel capture, and scan-out strobe. You can add initialization vectors at the start of the test program using the **read_init_protocol** command. Shown below is the default protocol inferred by TC for a design after scan-insertion.

```
test_protocol() {
  period : 100.00 ;/* test_default_period */
  delay : 5.00 ;/* test_default_delay */
  bidir_delay : 55.00 ;/* test_default_bidir_delay */
  strobe : 95.00 ;/* test_defaut_strobe */
  strobe_width : 0.00 ;/* test_default_strobe_width */

  clock() {
    period : 100.00 ;
    waveform : {45.00, 55.00} ;/* create_test_clock */
    sources : clk ;
  }
  protocol_start() {
    foreach_program() {
/* INITIALIZATION VECTORS */
      vector() {
        set(all_ports,"0,[7]X,[4]M");
      }
      vector() {
        set(all_ports,"0,[4]X,1,[2]X,[4]M");
      }
/* You can make changes above this portion without requiring a Custom Test
                                Protocol license */

      foreach_pattern() {

/* SCAN IN-OUT vectors */
        stream(101) {
          set(all_ports,"C,[4]U,[2]1,Si,M,So,[2]M");
        }
/* Parallel Measure vector */
        vector() {
          set(all_ports,"0,[4]Pi,1,[2]Pi,[3]Po,Pio");
        }
/* Parallel Capture Cycle */
        vector() {
```

```
          set(all_ports,"Cp,[4]U,1,[2]U,[4]M");
        }
  /* Scan Out strobe */
        vector() {
          set(all_ports,"0,[4]U,[2]1,U,M,So,[2]M");
        }
      }
    }
  }
```

# 8.6 Test Compiler Tips

In this section we summarize the different points that one must consider when incorporating testability into the design.

- Perform a testability analysis at the sub-module level to isolate local testability issues. Also, run ATPG on a small sample fault list to detect any ATPG conflict situations as shown below:

  ```
  create_test_patterns -sample 3
  ```

- Set the timing attributes for scan testing of the design appropriately, to avoid simulation mismatches. Set **test_default_bidir_delay** to a value after the rising edge of the capture clock. This variable specifies the time when data is applied to the bi-directionals in the input mode.
- The scan-in and scan-out ports must be shared with functional ports only if supported by the ASIC vendor. When choosing functional ports to share, remember that sharing functional ports for scan-in ports increases loading on the ports. Similarly, sharing scan-out ports results in additional delay through a mux. Select functional ports which do not have very tight timing requirements.
- Avoid potential illegal paths and diverging scan chains. For a Return-To-Zero clock, ensure that all negative edge triggered flops come before the positive-edge triggered flops in the same scan chain. Similarly, for a Return-To-One clock, all positive edge triggered flops should come before negative edge-triggered flops.
- During functional mode timing analysis, there might be multi-cycle paths between the flip-flops. In the test-mode, all the flip-flops are clocked in the same cycle. Hence, perform timing analysis on the complete design after scan-insertion, without any path exceptions. Also, use the **set_clock_skew** command to account for all the delays on the different clock branches. If the clock tree is in place, use the **set_clock_skew -propagated** command.

## 8.7 Examples Showing the Entire Test Synthesis Flow

In this section, we summarize the entire test synthesis flow by means of an example. Example 8.2 outlines the methodology to be followed for a design with bidirectional ports, three-state O/P ports, I/P and O/P ports. We illustrate the entire flow ranging from synthesis to test insertion, and JTAG synthesis followed by pad synthesis (assuming of course, that the ASIC vendor library supports that).

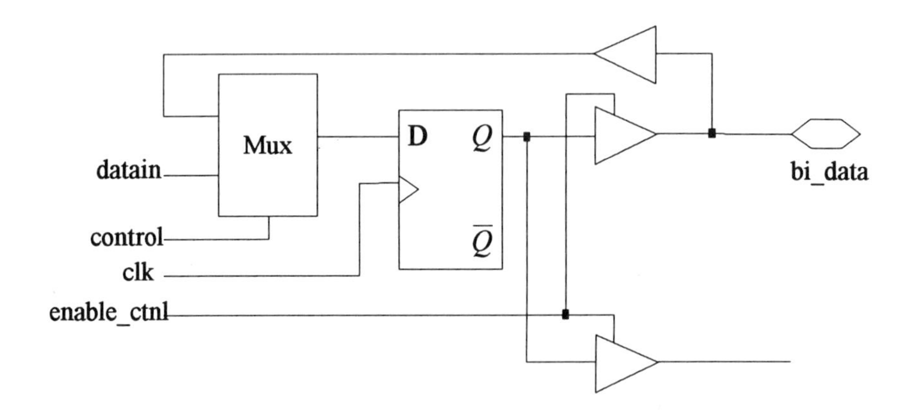


**Figure 8.9 Design Showing Bidirectional Outputs**

### Example 8.2 HDL Code for a design with bi-directional ports

```
library IEEE;
use IEEE.std_logic_1164.all;

entity bidir is
  port (clk, control, enable_ctrl, datain: in std_logic;
 bi_data: inout std_logic;
 tri_data: out std_logic);
end bidir;
architecture behv of bidir is
signal temp: std_logic;
begin
  process
  begin
wait until clk = '1' and clk'event;
if (control = '1') then
  temp <= datain;
```

```
else
  temp <= bi_data;
end if;
  end process;
  bi_data <= temp when (enable_ctrl = '0') else
    'Z';
  tri_data <= temp when (enable_ctrl = '1') else
    'Z';
end behv;

/* dc_shell script */
read -f vhdl bidir.vhd
set_test_methodology full_scan
set_scan_style multiplexed_flip_flop
compile
check_test
insert_scan
check_test
create_test_patterns -sample 5
group -design_name Core -cell_name core1 filter(find(cell "*"),"(@is_combinational
                              == true) || (@is_sequential == true) || (@is_hierarchical
                              == true) || (@is_black_box == true)")
current_design bidir
insert_jtag -no_pads
check_test
create_test_patterns
write_test -format vhdl
```

# 8.8 Classic Scenarios

**Scenario 1**

How does one query the library (db file) in DC for information about pins, cells, and functionality for library cells.

**Solution**

To check if a particular output pin of a cell in a library libA has a function attribute, use the get_attribute command as shown below:

```
dc_shell> get_attribute find(pin, libA/FD2SP/QN) function
Performing get_attribute on port 'QN'.
{"IQN"}
```

This shows that the QN pin is used in the normal functional mode. The command find(pin, libA/FD2/*) can be used to find all pins of a cell. The attribute direction shows the direction of the pins. Also, use the command list -libraries to find the internal name of the technology library you are using.

To find all muxed flip-flops cells in the library libA, use the following dc_shell script. For other scan styles this script can be easily modified. The file scan.cells contains a list of all the scan cells in the library and their pins.

```
/*dc_shell script*/
sh rm scan.cells
message = ""
filter find(cell, libA/*) "@is_sequential == true";
seq_list = dc_shell_status;
foreach(cell_name, seq_list){
  filter find(pin, cell_name + "/*") "@signal_type == test_scan_in"
  if (dc_shell_status != {}){
      message = cell_name
      message = "Found a scan cell " + message
      echo message >> scan.cells
      pin_name = find(pin, cell_name + "/*");
      echo pin_name >> scan.cells
  }
}
Sample output on a library, libA, gives the following:
Found a scan cell libA/FD2SP
{D, CP, CD, TI, TE, Q, QN}
Found a scan cell  libA/FJK2SP
{J, K, CP, CD, TI, TE, Q, QN}
Found a scan cell libA/FD3SP
{D, CP, CD, SD, TI, TE, Q, QN}
Found a scan cell libA/FJK3SP
{J, K, CP, CD, SD, TI, TE, Q, QN}
```

**Scenario 2**

You have a design with scan inserted. The scan style used is multiplexed flip-flop and methodology, full scan. Running check_test on the post scan design gives a clean report

```
Information: Inferred capture clock group : CLK. (TEST-262)
  ...binding scan-out state...
  ...simulating serial scan-out...
  ...data scanned-out from 10 cells (total scan-out 11)...
  ...simulating parallel vector...
```

```
Information: There are 11 scannable sequential cells. (TEST-295)
Information: Test design rule checking completed with 0 warning(s)
and 0 error(s). (TEST-123)
1
```

After formatting the vectors generated from ATPG, you observe that in the patterns, TC does not strobe the scan-out during scan-shift in some patterns, for one scan cell. Shown below is a segment of the formatted Verilog vectors which shows that the scan-out strobe is missing at time 3955.

```
#10 ; /* 3805 */
assign _si=1'b1;
 #40 ; /* 3845 */
assign _ck=1'b1;
 #10 ; /* 3855 */
assign _ck=1'b0;
 #40 ; /* 3895 */
assign _e_so=1'b1;
-> _check_so;
 #50 ; /* 3945 */
assign _ck=1'b1;
 #10 ; /* 3955 */  <-------- Scan-out strobe missing in this cycle
assign _ck=1'b0;
 #90 ; /* 4045 */
assign _ck=1'b1;
 #10 ; /* 4055 */
assign _ck=1'b0;
 #50 ; /* 4105 */
assign _pi=4'b0010;
assign _si=1'b0;
 #90 ; /* 4195 */
assign _e_so=1'b0;
-> _check_so;
```

**Solution**

TC will not strobe data during scan-out for a particular scan cell if it captures an 'X' during the capture cycle. This could be due to any test design rule violation, a test_isolate attribute, a black-box in the transitive fan-in of the data pin of the corresponding scan cell, or a combinational feedback loop. In those test patterns when 'X' is sensitized to the data pin of the scan cell, TC will not strobe the scan out for the corresponding scan cell. For example, consider an OR gate feeding the D pin of a scan cell, one input of the OR gate being an 'X' (unknown), while the other input is driven by a primary input. For the patterns where the primary input is a '1', TC will mask the 'X' and when the primary input is a '0', TC will sensitize the 'X' to the data pin.

**Scenario 3**

You have gone through the recommended flow for test synthesis and have generated test vectors. The **check_test** results on the post-scan design do not show any major warnings or errors. When simulating the test vectors, you get simulation mismatches while simulating the scan check vectors, and during the scan-out cycles of the ATPG vectors. Why?

**Solution**

If the mismatch is during simulation of the scan check vectors, perform a min-delay timing analysis on your design to identify any hold violations in the scan path. Hold violations in the scan path could cause mismatches. If the mismatches are during the scan-out phase of the ATPG vectors, check if the mismatches are between a known expected 1/0 and an 'X' (unknown) response or a known 0/1 response. If the response is an 'X', there are most likely hold violations in the datapath. Run a min-delay timing analysis to narrow down the paths with hold violations, and fix those violations. If the mismatches are between known values, verify if any **set_test_assume** has been used during ATPG (**report_test -assertions** will show them). Note that TC does not verify the values specified by the **set_test_assume** command. Also make sure that the synthesis models and simulation models are consistent. An example of this could be using a structural netlist for a ASIC vendor hard macro cell which does not match with the simulation model provided by the vendor.

**Scenario 4**

You execute **insert_scan** at the core level of your design followed by **check_test**. TC report no errors. Then you create another hierarchy around your core design and manually instantiate pads. On executing **check_test** at the top level of the design with the pads included in the netlist, TC reports, "No scan-path found".

**Solution**

This occurs when **set_test_methodology -existing_scan** has not been specified from the top level of the design. Also, ensure that the **signal_type** attributes for all the scan ports have been specified along with any required test assertions (example **set_test_hold**), and execute the **check_test** command from the top level.

**Scenario 5**

You are running DRC (**check_test**) on the top level after integrating your ASIC. Finding that TC is inferring the asynchronous reset line as a clock and all the scan cells with asynchronous reset pins being classified as constant-logic black-box cells.

**Solution**

TC infers a clock by backtracking from pins which have setup and hold arcs specified. If the reset signal is connected to either enable pins of latches (which might be classified as black boxes), synchronous RAM pins, or any pins which have setup and hold arcs specified in the library, it shall be inferred as a clock by TC. If these cells are already classified as black-boxes, specify a **test_isolate** attribute on them using the **set_test_isolate** command. This will prevent the reset line from being inferred as a clock.

**Scenario 6**

You are attempting to insert an approximately 70 cell scan chain deep within the hierarchy of the chip. The hierarchy is shown in Figure 8.10.

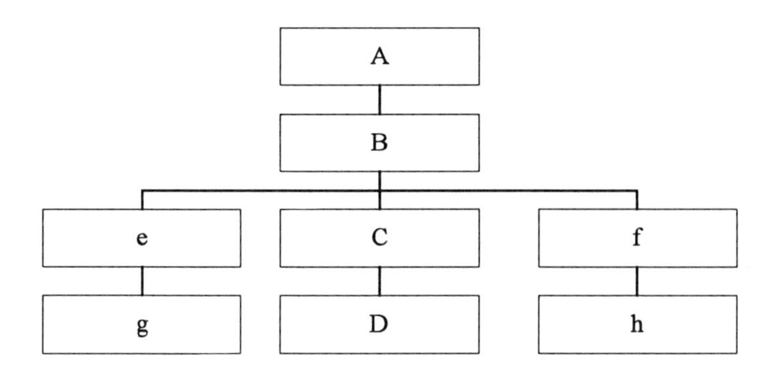


**Figure 8.10 Hierarchy of Design**

Design D contains the 70 cells you wish to scan. When you set the current design to D and generate vectors, it appears that TC naturally has control over all the primary inputs to D and observability of Ds outputs. This, however, is not the case. The only access TC has to inputs and outputs of D, is through a scan chain that was inserted into D. This scan chain has it's serial input through a port in design A, as well as a serial-out, also in design A.

What should the current design be set to in order to generate these vectors? The two possible choices are A and D. If you set current design to D, TC thinks it has access to inputs that it my not have. If you set current design to A, TC has to create vectors for a design buried 4 or 5 levels deep into the hierarchy. The scan style used is the multiplexed flip-flop scan style.

**Solution**

Using the actual tester you will be applying patterns at level A. Hence, your test patterns should be using the ports of top level, A. At level A you identify the scan-in, scan-out and scan-enable ports and TC should recognize the scan chain through block D. All the other non-scan sequential cells in your design will be classified as back-boxes and should be X generators. Then, you can run ATPG from level A and the tester will generate vectors and control the primary inputs (PIs) at level A, to generate patterns for the complete hierarchy. If you are only interested in finding the coverage of block D, set the test_dont_fault attribute on the other blocks.

**Scenario 7**

You have inserted scan in your design using the insert_scan command. Your scan style is multiplexed flip-flop and the test methodology, full scan. Before running check_test you associate a signal_type attribute test_scan_clock with the clock port of your design. Why does check_test gives the following output?

```
  ...2046 bits scanned-in to 2046 cells (total scan-in 2046)...
  ...simulating parallel vector...
  ...binding scan-in state...
  ...simulating parallel vector...
  ...simulating parallel vector...
  ...creating capture clock groups...
Warning: Data can not be captured into cell D312 (SDF1). (TEST-310)
Information: There are 2045 other cells with the same violation. (TEST-171)
  ...binding scan-out state...
  ...simulating serial scan-out...
  ...simulating parallel vector...
Information: There are 2046 scan controllable only sequential cells. (TEST-295)
Information: There are 12 black-box sequential cells. (TEST-295)
Information: Test design rule checking completed with 3 warning(s) and 0 error(s).
                              (TEST-123)
```

**Solution**

For multiplexed flip-flop scan style, the functional clocks are used in the test mode. A common mistake is to specify set_signal_type test_scan_clock for the system clocks in a multiplexed flip-flop design. The signal_type attribute test_scan_clock should only be used to identify the test clock port for clocked scan designs. When this attribute is placed on system clocks in a multiplexed flip-flop design, it has the side effect of holding the system clocks inactive during the parallel capture cycle. Test design rule checking will indicate that data cannot be captured into the sequential cells, and all otherwise scannable cells are classified as scan controllable only. Do not associate the signal_type attribute, test_scan_clock, with the clock port. If it already exists, remove it using this command remove_attribute command.

**Scenario 8**

You read in an ASCII VHDL netlist for your ASIC with the JTAG logic included into TC. The scan chain specific information such as scan-in, scan-out, scan-enable are specified along with other test mode assertion conditions. How to go about handling the included JTAG logic in the design?

**Solution**

TC does not generate test patterns for the JTAG logic. This is because the JTAG logic have many test design rules violations associated with it. The JTAG logic is put in the transparent mode throughout ATPG. This is easily achieved when the synthesized JTAG logic includes the optional asynchronous reset port. According to the IEEE 1149.1 JTAG specs activating the JTAG TRST port asynchronously puts the TAP controller into the Idle/Reset state. In this state, all the BSRs are put in the feedthrough mode, thereby making the JTAG logic completely transparent for ATPG.

If you do not synthesize the asynchronous JTAG TRST port, an initialization protocol is needed to put the TAP controller in the Idle/Reset state. This is accomplished by holding the JTAG port TMS high and clocking the JTAG clock TCK for five cycles. The TAP controller state machine is designed such that this always puts it in the Idle/Reset state. The following TC commands are needed to identify the JTAG logic for TC:

```
current_design top_level
set_attribute jtag_inserted design_name -type boolean true
set_signal_type /* to identify all the JTAG ports */
```

**Scenario 9**

You are using LSI Logic as your ASIC vendor. How is Test Compiler's default test protocol different from that of LSI Logic in terms of handling bidirectional ports?

**Solution**

LSI Logic has a four cycle test application sequence which imposes certain requirements on bidirectionals and their mode in each cycle. All bidirectional ports must be in the input mode in cycles 1,3 and 4. This is accomplished by activating the *bidirectional output inhibit signal* in the cycles 1,3 and 4. This signal that controls the direction of the bidirectional ports needs to be a primary input port and must have the signal_type attribute test_bidir_control[_inverted] in TC.

In cycle 2, the bidirectional output inhibit signal is not active and hence the ATPG interprets the direction of the bidirectional ports in this cycle.

The LSI test protocol has the following four cycles:

- Scan Shift

  The *bidirectional output inhibit* input port is active and all the bidirectional ports are inputs.

- Parallel Measure

  In this cycle the output inhibit port is inactive. LSI has bidirectional pads with pins TN and EN. TN is the pin to which the *bidirectional output inhibit* signal is connected to. When TN is active or 1, the bidirectional is in the input mode. When TN is inactive or 0, ATPG configures the bidirectional port direction through EN.

- Dead Cycle

  The *bidirectional output inhibit* input port is active and all the bidirectional ports are inputs.

- Parallel Capture

  The *bidirectional output inhibit* input port is active and all the bidirectional ports are inputs.

**Scenario 10**

How does the fault simulation done by ATPG software like TC differ from that of other fault simulators available?

**Solution**

During create_test_patterns, that is, ATPG Test Compiler internally simulates the test patterns using a 0 delay model. Other fault simulators like TestSim or Zycad do a full timing fault simulation. Also TC as well as other Test Synthesis tools account for faults detected in the parallel measure cycle and the parallel capture cycle. In general, faults may also be detected while shifting in a pattern into the scan chain. When using TestSim or other fault simulators these additional faults can be detected.

**Scenario 11**

In my design, the core logic block has 98% fault coverage. When I instantiate the core login in a top level and add I/O pads, the fault coverage goes down 96%. The top level has four 16 bit bi-directional ports. We need to bring up fault coverage to 98%.

**Solution**

The untested faults are most likely the faults associated with the enabling logic of the bidirectionals. Any fault at the enable pin of the three-state cell will cause a good machine (or faulty machine) value of Z on output of the three-state cell. This will cause a float condition (either in good machine or in the faulty machine), which must certainly be avoided. One solution could be bringing the control (tri-state) pin of the bidirectional ports to an output port.

**Scenario 12**

You have performed JTAG synthesis using TC and turned off synthesis of the JTAG asynchronous reset port. When doing a VHDL functional gate-level simulation of your ASIC you are trying to initialize the TAP controller to the Idle/Reset state by holding TMS high and clocking JTAG TCK five times. But in your gate-level simulation the state of the TAP controller does not change from U (uninitialized).

**Solution**

At the start of simulation, all the sequential elements on the ASIC including those in the TAP controller are uninitialized. Since the TAP controller does not have a reset pin (either synchronous or asynchronous) the "U"s keeps propagating to the data pin of the flops thereby not being able to move the TAP controller to the *Idle* state. One way is to arbitrarily initialize the state vector of the TAP controller to a known value at the start of simulation and then simulate your regular functional vectors. If you are using Synopsys VHDL System Simulator for your gate-level simulation, the VSS commands `hold` and `assign` can be used to clock a known arbitrary value into the TAP controller.

**Scenario 13**

You have a design where the asynchronous pins of your flip-flops are controlled by an active low top level primary input RESET port. The stuck-at-one faults on your reset line are reported as untestable. Why?

**Solution**

Check if the Synopsys variable `atpg_test_asynchronous_pins` has been set to false. When true (the default value), both stuck-at-0 and stuck-at-1 faults on asynchronous pins are considered for test generation. When set to false, it allows the user to force asynchronous input pins to all flip-flops to inactive states during the parallel measure and capture phases of a scan test. This implies that no tests will be generated for stuck-at-inactive-value faults on the asynchronous pins and those faults will be reported as untestable.

**Scenario 14**

You have generated test patterns for a design and formatted the vectors in the WGL format. The expected response for the first scan cell in the scan chain is always 'X' (not strobed). Why? Shown below is the relevant section from the patterns.

```
scanstate
```

```
 state0 := SC1_G(1001);
 state1 := SC1_G(0111);
estate1 := SC1_G(X111);
 state2 := SC1_G(1010);
estate2 := SC1_G(X000);
...
...
end
```

**Solution**

In the default test protocol inferred by TC, the last scan cell in the scan chain is strobed in the scan-out strobe cycle after the parallel capture cycle. If this is not the case, the next scan-shift cycle will overwrite the captured response in the last scan cell before it is strobed. Hence in the last cycle of the scan shift there is no expected response, that is an X.

**Scenario 15**

In the default test protocol inferred by TC there is a separate scan-out strobe cycle after the parallel-capture cycle. Why is this cycle not merged with the parallel-capture cycle?

**Solution**

In designs where a functional port is shared with a scan-out port, TC synthesizes a mux to select between the last scan cell in the scan chain and the functional logic. The *scan-enable* signal is connected to the select line of the mux such that when the *scan-enable* is active, the last scan cell is selected. Since the scan-enable is generally inactive in most patterns to capture faults on the data path, a separate scan-out strobe cycle is required.

**Scenario 16**

How does Test Compiler configure the JTAG logic on my ASIC during ATPG.

**Solution**

Test Compiler configures the JTAG logic in the feedthrough mode during Automatic Test Pattern Generation (ATPG). This can be achieved by either activating the asynchronous JTAG reset port (if present) or by clocking the TAP controller for 5 cycles and holding TMS high. Both these approaches place the TAP controller in the "Idle/Reset" state in which the JTAG logic remains in the feedthrough mode.

## Recommended Further Readings

1. Test Compiler Reference Manual
2. Test Compiler Streamlined Methodology Application Note
3. Synopsys Newsletter "Impact" Support Center Q&A

CHAPTER 9

# *Interfacing Between CAD Tools*

For almost every phase of the design cycle there are different CAD tools available. Schematic capture, simulation, synthesis, verification, PLD optimization, datapath synthesis, test synthesis, floor planning and layout tools are all required to work together in a seamless fashion to achieve maximum productivity. The CAD Framework Initiative (CFI) was primarily undertaken to improve tool inter-operability and establish certain standards. Current widely accepted standards such as EDIF for netlists and schematics (not to mention the different available flavors of EDIF), the Standard Delay Format (SDF) for back annotated delays and the Phyiscal Data Exchange Format (PDEF) for physical cluster information are examples of existing de facto standards.

In this chapter, we discuss both SDF and EDIF formats. The discussion of these formats has been provided to help the reader better deal with issues related to interfacing between CAD tools. A listing of all the currently available input and output formats to and from DC is also provided.

## 9.1 Electronic Data Interchange Format (EDIF)

EDIF files are output formats written by EDA tools for other EDA tools. This fact has led to a popular misconception that EDIF files are humanly incomprehensible. However, a little research will reveal that EDIF files are, in fact, not as complicated as they appear to be. Most often, when going from one EDA tool to another via EDIF, a backslash or an underscore in the wrong place, for example, can be very frustrating and time consuming. A better understanding of the intricacies of EDIF will help in overcoming these delays. This section provides an understanding of an EDIF schematic generated by DC. By default, DC writes out a schematic in EDIF. In other words, the variable

```
edifout_netlist_only = false        /* by default */
```

To write out an EDIF netlist from DC instead of an EDIF schematic, set this variable to true prior to writing out EDIF. Also, you must create a schematic of the design using the create_schematic command before writing out a schematic EDIF. The following two dc_shell commands must be executed to write out a design in schematic EDIF format from DC.

```
current_design = TOP
create_schematic -hierarchy
write -f edif -o correct.edif -hierarchy
```

Although EDIF is a standard, most EDA tools have their own flavor of EDIF. Some design houses have edif2edif translators! To overcome the problems due to different flavors of EDIF, most CAD tools have several variables which determine the flavor of the EDIF written out. That is, depending on the target tool, a set of variables need to be specified when writing out EDIF from DC. Similarly, when reading in an EDIF generated by other tools, it might be required that certain variables be specified.

Before getting into greater detail, first, a few basic definitions:

- In EDIF, every construct consists of an open parenthesis, a keyword, the body of the construct, and a close parenthesis. The body of the construct can include other constructs.
- In EDIF, a string is a sequence of characters within quotes. An identifier is the name of an entity.
- In an EDIF netlist description, entities include ports, nets, cells, and libraries. The syntax for an identifier requires that it begins with a letter, and be composed of letters, numbers, and underscore characters. An identifier that does not begin with a letter may begin with an ampersand, which is ignored.

Example 9.1 is a design made of just one cell (a latch) as shown in Figure 9-1. The design (say, latch1) has been mapped to the lsi_10k technology library. Believe it or not, the EDIF file for this design has a total of 206 lines. Brevity is the soul of wit? Not as far as EDIF is concerned!

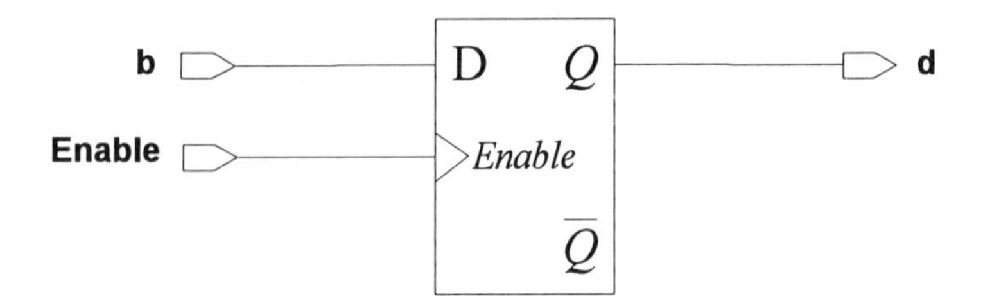


**Figure 9.1 Design Latch1 with Instantiated Latch LD1**

## Example 9.1 EDIF File

What follows is the EDIF file interspersed with brief descriptions. It has been shortened for clarity and easy understanding.

First the default name, Synopsys_edif is declared. This can be replaced by the actual design name by setting the edifout_design_name variable. This is followed by the (external...) construct, which provides information about the technology library used.

```
edif Synopsys_edif (edifVersion 2 0 0) (edifLevel 0)
 (keywordMap (keywordLevel 0)) (status)
 (external (rename lsi_10k_sdb "lsi_10k.sdb") (edifLevel 0)
 (technology (numberDefinition (scale 1 (e 2480469 -12) (unit DISTANCE)))
```

This is followed by information about different layers of the schematic, for example:

```
  (figureGroup default) (figureGroup cell_name_layer (color 100 50 0))
  )
```

Then, the different cells of the external library that are used in this design are described partially. In other words, the (cell...) construct includes an (interface ...) construct but no contents construct. For example, the cell LD1 of the lsi_10k library is shown. The interface construct includes information about the ports, their properties and their directions (INPUT, OUTPUT)

```
 (cell LD1 (cellType GENERIC)
   (view Schematic_representation (viewType SCHEMATIC)
    (interface (port D (direction INPUT)) (port G (direction INPUT))
     (port Q (direction OUTPUT)) (port QN (direction OUTPUT))
     (port VDD (direction INPUT)
      (property implicitPortClass (string "VDD") (owner "Schematic_TSC"))
      (property portType (string "supply") (owner "Schematic_TSC"))
      (property supplyType (string "power") (owner "Schematic_TSC"))
     )
     (port GND (direction INPUT)
      (property implicitPortClass (string "GND") (owner "Schematic_TSC"))
      (property portType (string "supply") (owner "Schematic_TSC"))
      (property supplyType (string "digitalGround") (owner "Schematic_TSC"))
     )
    )
   )
  )
 )
```

This is followed by the (library...) construct. This groups cells of the external construct. It differs from the (external...) construct in that the components defined are fully described. In other words, they contain a (cell...) construct, an (interface...) construct, and a (contents...) construct. The (contents...) construct further consists of an (instance...) construct and a (net...) construct.

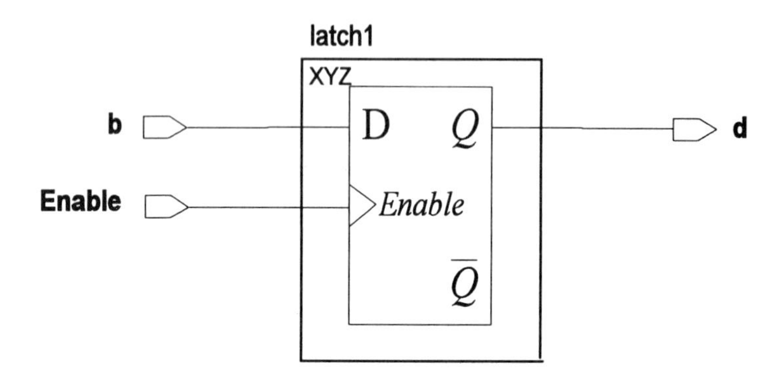


**Figure 9.2 Design with an Additional Level of Hierarchy**

Example 9.2 is identical to Example 9.1 except that there exists an additional level of hierarchy above the latch, called XYZ. If this contains the latch and say some additional logic, the library construct will have a cell XYZ construct, followed by (interface...) construct defining the ports of this block, and the cells LD1 and the gates will appear under the (contents...) construct.

## Example 9.2 EDIF File Showing Hierarchy

```
(library DESIGNS (edifLevel 0)
 (technology (numberDefinition (scale 1 (e 2480469 -12) (unit DISTANCE)))
  (figureGroup default) (figureGroup cell_name_layer (color 100 50 0))
 )
 (cell latch1 (cellType GENERIC)
  (view Schematic_representation (viewType SCHEMATIC)
   (interface (port clk (direction INPUT)) (port din (direction INPUT))
    (port lat (direction OUTPUT))
    (port VDD (direction INPUT)
     (property implicitPortClass (string "VDD") (owner "Schematic_TSC"))
     (property portType (string "supply") (owner "Schematic_TSC"))
     (property supplyType (string "power") (owner "Schematic_TSC"))
    )
```

```
(port GND (direction INPUT)
 (property implicitPortClass (string "GND") (owner "Schematic_TSC"))
 (property portType (string "supply") (owner "Schematic_TSC"))
 (property supplyType (string "digitalGround") (owner "Schematic_TSC"))
)
(symbol (boundingBox (rectangle (pt -3072 -2048) (pt 3072 2048)))
 (portImplementation clk
  (connectLocation (figure cell_layer (dot (pt -3072 1024))))
 )
 (portImplementation din
  (connectLocation (figure cell_layer (dot (pt -3072 -1024))))
 )
 (portImplementation lat
  (connectLocation (figure cell_layer (dot (pt 3072 0))))
 )
)
)
(contents
 (page &1 (pageSize (rectangle (pt -89856 -7680) (pt 6144 13568)))
  (portImplementation clk
   (connectLocation (figure port_layer (dot (pt -16384 4096))))
   (figure port_layer (path (pointList (pt -18944 3456) (pt -18944 4736)))
   )
   (figure port_layer (path (pointList (pt -18944 4736) (pt -17664 4736)))
   )
   (figure port_layer (path (pointList (pt -18944 3456) (pt -17664 3456)))
   )
   (figure port_layer (path (pointList (pt -17664 3456) (pt -16384 4096)))
   )
   (figure port_layer (path (pointList (pt -17664 4736) (pt -16384 4096)))
   )
  )
```

All port implementations are described as shown above. The other port implementations have not been shown. Below the *(content...)* construct is continued, in order to show the *(instance ..)* and *(net...)* constructs.

```
  (instance lat_reg
   (viewRef Schematic_representation
    (cellRef LD1 (libraryRef lsi_10k_sdb))
   )
   (transform (origin (pt -14336 3072)))
  )
  (net lat16 (joined (portRef din) (portRef D (instanceRef lat_reg)))
```

```
      (figure net_layer
       (path (pointList (pt -16384 12288) (pt -14336 12288)))
      )
     )
     (net lat (joined (portRef lat) (portRef Q (instanceRef lat_reg)))
      (figure net_layer (path (pointList (pt -8192 12288) (pt -5120 12288))))
     )
     (net n17 (joined (portRef clk) (portRef G (instanceRef lat_reg)))
      (figure net_layer (path (pointList (pt -16384 4096) (pt -14336 4096))))
     )
     (net VDD (joined (portRef VDD) (portRef VDD (instanceRef lat_reg))))
     (net GND (joined (portRef GND) (portRef GND (instanceRef lat_reg))))
    )
   )
  )
 )
)
(design Synopsys_edif (cellRef latch1 (libraryRef DESIGNS)))
)
```

Finally, latch1 is the top-level component of the design.

## 9.2 Forward and Back-annotation

Forward annotation is the process of providing timing constraints and physical cluster information to back-end tools, and timing delays to gate-level simulation tools. Back-annotation on the other hand is the process by which resistance, capacitance, phyical cluster and delay information after place and route are provided to DC. The set_load and set_resistance commands can be used to back annotate capacitance and resistance values. The Standard Delay Format (SDF) is a de facto standard for annotating delay values. The Phyical Data Exchange Format (PDEF) is a means for transferring phyical cluster information between Synopsys DC and back-end tools like place and route and floorplanning tools.

### 9.2.1 Standard Delay Format (SDF)

SDF is an industry standard for communicating timing information between EDA tools. Though the SDF language supports a wide range of constructs; only a few are supported for synthesis. DC can both read in SDF as well as output SDF. There are two scenarios in which one might write out SDF from Synopsys, namely, during forward annotation of constraint information from DC to back-end tools or during forward annotation of timing information from DC to simulation tools for gate-level simulation. While the former is done using write_constraints command, the latter is done using the write_timing command. Shown below are the steps to write SDF from DC after one has initially read in the source VHDL.

```
compile
change_names -rules vhdl
write -f vhdl current_design -hier -output netlist.vhd
write_timing -format SDF
write_constraints -format SDF
```

One can back-annotate timing information using SDF from back-end tools such as floorplanning and place and route tools. After the floorplanning phase is complete, and before the design is passed on to physical layout (place and route), the design's timing behavior can be verified once more within the synthesis environment. This time, the more accurate net delays, and cell delays are used in place of the values estimated by DC. The estimates (provided by a floorplanner) for the net delays, and cell delays can be back annotated into DC using the read_timing command as shown below.

```
read -format db design.db /* read in hierarchical db of design */
current_design = top /* set the current design to top level of hierarchy */
read_timing -format SDF delays.sdf /* read SDF delay file */
```

DC supports the following SDF timing constructs for forward annotation to simulation tools and back-annotation from back-end tools:

- INTERCONNECT and PORT for net delay
- IOPATH for cell delay
- SETUP, HOLD, and SETUPHOLD for timing checks.

The IOPATH construct specifies the cell intrinsic delay between input pins and output pin. For sequential elements, the setup and hold timing arcs are specified by the TIMINGCHECK group in the SDF file. Example 9-3 shows an example of the IOPATH and the TIMINGCHECK constructs in an SDF file.

### Example 9.3 TIMINGCHECK in SDF File

```
(CELL (CELLTYPE "dfntnb")
 (INSTANCE I1.int_count_reg_3)
 (DELAY (ABSOLUTE
  (IOPATH CP Q (1.144:1.144:1.144) (.797:.797:.797))
  (IOPATH CP QN (.714:.714:.714) (.794:.794:.794))
 ))
```

```
(TIMINGCHECK
 (SETUP D (posedge CP) (.280))
 (HOLD D (posedge CP) (.000))
 (WIDTH (posedge CP) (.300))
 (WIDTH (negedge CP) (.350))
)
)
```

DC uses the INTERCONNECT construct to describe net delays in the SDF file written out by write_timing. When reading in an SDF file, if the PORT construct is used for net delays, it is first converted to an INTERCONNECT construct and then annotated to the design. The INTERCONNECT construct is used to specify the net connect delay between two pins. The floorplanning or P&R tool may choose to lump the net transition delay either in the INTERCONNECT or IOPATH construct. Example 9.4 shows the specification of the net delay between the source pin U11/Z and the destination pin I2.int_count_reg_3/CP using the INTERCONNECT construct. The SDF file shows three values one each for minimum, typical, maximum timing delays. Only one value of these can be read from the SDF file. Users may specify the value (minimum, typical, maximum) that should be read by DC using the following dc_shell variables:

```
sdfin_fall_net_delay_type = maximum
sdfin_rise_net_delay_type = maximum
sdfin_fall_cell_delay_type = maximum
sdfin_rise_cell_delay_type = maximum
```

## Example 9.4 INTERCONNECT in SDF File

```
(CELL (CELLTYPE "top")
(INSTANCE )
(DELAY (ABSOLUTE
 (INTERCONNECT U11.Z I2.int_count_reg_3.CP (.050:.050:.050) (.052:.052:.052))
 (INTERCONNECT U11.Z I2.int_count_reg_1.CP (.078:.078:.078) (.081:.081:.081))
 (INTERCONNECT U11.Z I2.int_count_reg_0.CP (.089:.089:.089) (.092:.092:.092))
```

When forward annotating timing constraints to timing driven back-end tools using the write_constraints command, the DC writes out the PATHCONSTRAINT construct in the SDF file as shown in Example 9.5. This construct specifies the required delay on a path. In our example SDF file, the timing requirement on the path from the clock pin of I2/int_count_reg_0 to the data pin of I2/int_count_reg_1 is 4.696.

**Example 9.5 PATHCONSTRAINT in SDF File**

```
(DELAYFILE
(SDFVERSION "OVI 1.0")
(DESIGN "top")
(DATE "Sat Mar 16 11:07:33 1996")
(VENDOR "custom")
(PROGRAM "Synopsys Design Compiler cmos")
(VERSION "v3.4a")
(DIVIDER /)
(VOLTAGE 4.75:4.75:4.75)
(PROCESS)
(TEMPERATURE 70.00:70.00:70.00)
(TIMESCALE 1ns)
(CELL
 (CELLTYPE "top")
  (INSTANCE)
  (TIMINGCHECK
   (PATHCONSTRAINT I1/int_count_reg_0/cp I1/int_count_reg_0/q I1/U7/a2 I1/U7/zn
                   I1/U10/b1 I1/U10/zn I1/U11/a1 I1/U11/zn
                   I1/int_count_reg_3/d (4.619:4.619:4.619) )
...
...
 (PATHCONSTRAINT I2/int_count_reg_0/cp I2/int_count_reg_0/q I2/U9/a2 I2/U9/z
                 I2/int_count_reg_1/d (4.696:4.696:4.696) )
  )
)
)
```

## 9.2.2 Physical Design Exchange Format (PDEF)

PDEF is an industry standard format for representing physical hierarchy information. Generally designers partition their hierarchy based on logical functionality. Floorplanners and Placement and Route tools may need to change this hierarchy to be able to meet timing and to route the design. In such cases, the hierarchy defined during floorplanning does not have to exactly match the logical design hierarchy.

Example 9.6 shows a PDEF file. It defines three physical hierarchies (clusters) I2; top_ga, and I1. Also, the cell instances in each of the clusters is also defined. DC can read in this information and perform further optimization on the design using this physical hierarchy information.

### Example 9.6 PDEF File

```
(CLUSTERFILE
  (PDEFVERSION "1.0")
  (DESIGN "top")
  (DATE "Sat Mar 16 11:33:10 1996")
  (VENDOR "COMPASS")
  (PROGRAM "ChipPlanner-GA")
  (VERSION "v8r4.9.0")
  (DIVIDER /)
  (CLUSTER (NAME "I2")
    (UTILIZATION .4328)
    (MAX_UTILIZATION 100.0000)
    (CELL (NAME I2/int_count_reg_2))
    (CELL (NAME I2/int_count_reg_3))
    (CELL (NAME I2/U8))
    (CELL (NAME I2/U11))
    (CELL (NAME I2/U10))
    (CELL (NAME I2/U7))
    (CELL (NAME I2/int_count_reg_1))
    (CELL (NAME I2/int_count_reg_0))
    (CELL (NAME I2/U9))
  )
  (CLUSTER (NAME "top_ga")
    (UTILIZATION .4379)
    (MAX_UTILIZATION 100.0000)
    (CELL (NAME U11))
    (CELL (NAME TC_reg))
    (CELL (NAME U12))
  )
  (CLUSTER (NAME "I1")
    (UTILIZATION .4361)
    (MAX_UTILIZATION 100.0000)
    (CELL (NAME I1/int_count_reg_1))
    (CELL (NAME I1/int_count_reg_0))
    (CELL (NAME I1/U9))
    (CELL (NAME I1/U7))
    (CELL (NAME I1/U8))
    (CELL (NAME I1/U10))
    (CELL (NAME I1/int_count_reg_2))
    (CELL (NAME I1/int_count_reg_3))
    (CELL (NAME I1/U11))
  )
)
```

## 9.3 Design Compiler Input/Output Formats

### Table 9.1 DC Input/Output Formats and File Extensions

| Format | Description | File extension |
|---|---|---|
| db | Synopsys internal database | .db |
| edif | Electronic Design Interchange | .edif |
| lsi | LSI Logic Corporation (NDL) netlist | .net |
| equation | Synopsys equation | .fnc |
| mentor | Mentor NETED do | .neted |
| pla | Berkeley (Espresso) PLA | .pla |
| st | Synopsys State Table | .st |
| TDL | Tegas Design Language (TDL) netlist | .tdl |
| Verilog | Cadence Design Systems, Inc. | .v |
| VHDL | IEEE Standard VHDL | .vhd |
| SDF | Standard Delay Format (Cadence) | .SDF |
| XNF | Xilinx Netlist | .xnf |

## 9.4 Classic Scenarios

**Scenario 1**

You have written out a design in EDIF schematic format from DC. When reading this into another EDA vendor tool, you get the error message that pins are out of grid.

**Solution**

This problem is difficult to discuss without a very specific example. Instead we provide a general background of the issues involving pin spacing and scale factors. The SCALE factor in the symbol library is a pin-spacing factor. The pin-spacing for the different commonly used EDA tools are as follows

- 1/16' for Cadence
- 1/10' for Valid/Viewlogic
- 1/4' or 1/10' for Mentor

It is possible to generate a Symbol library in EDIF format from most EDA tools. From DC, one can write out EDIF of a symbol library. Similarly, one can read in an EDIF file of a symbol library and create a Synopsys symbol library (.sdb) file.

```
read_lib -f edif symbols.edif -o symbols.slib
```

When generating a symbol library source from EDIF, the EDIF symbol reader automatically calculates the SCALE factor to be the Greatest Common Divisor (GCD) of all the pin coordinates. This tool generated value must not be changed or this could be the beginning of your worst inter-operability nightmare. Also the ROUTE_GRID should always be 1024 in the Synopsys symbol library (.slib file). It should not be changed unless the intention is to merely see the schematics in DA for debugging purposes and not for writing out designs from DC. Also it is a requirement that the meter_scale attribute be consistent for all the symbol libraries used.

To transfer a design from DC to another EDA vendor tool via EDIF schematics, all the symbols used in the schematic must be from the ASIC vendor symbol library and not from the generic.sldb, the Synopsys generic symbol library. It is important to remember that a symbol library in system A is not the same as a symbol library in system B even if the cell names and interfaces match. Therefore, when dealing with schematic problems, the important issues to know are as follows:

- Which EDA tool was the symbol library generated in?
- What is the target EDA tool?
- What pin-spacing does the target-system use and what scale does the original symbol library have?

**Scenario 2**

You have synthesized your design using DC. You wish to write out an EDIF netlist. Depending on what your downstream tools accept, you wish to exercise control over the naming of power and ground cells when writing out EDIF. How can one go about controlling that in Synopsys.

**Solution**

There are three ways to represent power and ground when writing out EDIF from Synopsys. They can be represented as ports, cells, or nets. It is determined by the edifout_power_and_ground_representation variable.

1. Port representation: This is due to a cell construct in EDIF.

2. Net representation: Some ASIC vendors support nets as power and ground representation. The edifout_power_and_ground_representation variable must be set to a value net. Also, the following variables are used to identify the power and ground nets.

   edifout_power_net_name
   edifout_power_net_property_name
   edifout_power_net_property_value

3. Cell representation: In this case the power and ground are represented by cells from the ASIC vendor library. The edifout_power_and_ground_representation variable must be set to cell. The edifout_power_name variable identifies the cell and the edifout_power_pin_name identifies the pins on the power cell.

If there are power and ground cells in the symbol library, the schematic generator (create_schematic command) uses these cells in the schematic. When writing out EDIF, the EDIF writer filters out these cells if the power and ground representation is not cell. Also, if the edif writer does not find the power cells specified in the technology library, the power cells are included in the scope of each design in the edif output.

**Scenario 3**

DC assigns net names with *cell* in them. How can one avoid this?

**Solution**

Prior to running any compile steps in DC, use the define_name_rules command as shown below.

```
define_name_rules my_rules -type net -allowed "A-Z a-z _ 0-9 ()"
```

If you have already compiled your design, prior to writing out a netlist, use the following script. define_name_rules is a useful DC command that helps you define the names of nets and instances.

```
define_name_rules my_rules -type net -allowed "A-Z a-z _ 0-9 ()"
define_name_rules my_rules -replacement_char "_"
change_names -rules my_rules -hierarchy -verbose
```

**Scenario 4**

You have an SDF file generated after place and route where all the instance names are in upper case. But this is not so in the DC design database where all the instances are in lower case.

**Solution**

SDF is a case sensitive file. One option is to change instance names in the SDF file to lower case. Another option that can be exercised within DC, is to make all the instance names of the design upper-case. This can be done by using the commands define_name_rules and change_names. Here is an example:

```
define_name_rules upper_case -allowed "A-Z _ 0-9" -type cell
change_names -rules upper_case -hierarchy -verbose
```

**Scenario 5**

You have back annotated an SDF delay file which contained min/max/typ values into DC. But the timing reports generated by the report_timing command are identical for min or max timings.

**Solution**

DC reads only one set of values from the SDF file each time a file is back annotated. The following variables need to be set to indicate which set of values should be read:

- sdfin_fall_net_delay_type
- sdfin_rise_net_delay_type
- sdfin_rise_cell_delay_type
- sdfin_fall_cell_delay_type

These variables can be set to a value of minimum, maximum, or typical. To get different timing reports for the min and max values, back-annotation needs to be repeated for min or max values.

**Scenario 6**

When back annotating values into Synopsys, the wire load model is taken into account. If the values written out from the initial design also contain the WLM, is there a way to specify "no wire-load" model in DC?

**Solution**

It is likely that your technology library has a default_wire_load attribute. This attribute cannot be over-written by set_wire_load = "". Alternatively, create a dummy WLM with attributes set to 0.000. Then use this WLM in the design.

**Scenario 7**

When writing out EDIF from DC, DC issues the following error:

```
Error: "The meter scale in libraries 'generic.sdb' and 'yyy' aren't equal" (EDFO-2)
```

**Solution**

This error message occurs when writing out EDIF schematics if some of the cells used in the design are from the generic.sdb symbol library. DC automatically substitutes cells from this library if they are not available in the vendor's symbol library. The scales used in the generic symbol library and the vendor library are almost always different. One must determine which cells from the generic library are being used and ensure that symbols for those cells are available in the vendor symbol library.

**Scenario 8**

Are there any variables that one can specify which affect the naming style of buses or bused cells so that one can avoid the problem of "\" characters appearing in the Verilog netlist generated by DC?

**Solution**

The two variables that affect bus naming styles are as follows:

- bus_naming_style
- bus_dimension_separator_style

The second variable applies to cases where two dimensional arrays, that is, array of array are present in the design. By default, the bus_naming_style variable is set to %s[%d] and bus_dimension_separator_style is set to "]["

**Scenario 9**

You are generating FTGS VHDL simulation models from my Synopsys synthesis library. How does one turn off setup and hold checks on the scan path for multiplexed flipflop scan cells in the library when the scan_enable is inactive, that is, when the scan path is not selected.

**Solution**

The when and sdf_cond attributes in Library Compiler can be used to achieve this. Here is an example showing the relevant section of the library.

```
pin(TI) {
  direction : input;
  capacitance : 1;
  timing() {
    when : "TE";
    sdf_cond : "b == 1'b1" ;
    timing_type : setup_rising;
    intrinsic_rise : 1.3;
    intrinsic_fall : 1.3;
```

```
    related_pin : "CP";
  }
  timing() {
    when : "TE";
    sdf_cond : "b == 1'b1" ;
    timing_type : hold_rising;
    intrinsic_rise : 0.3;
    intrinsic_fall : 0.3;
    related_pin : "CP";
  }
}
```

**Scenario 10**

You are writing out an SDF file from DC and do not get any INTERCONNECT delay being written out even though you have specified the wire-load models.

**Solution**

The INTERCONNECT delay consists of the Connect Delay component and the Load Delay. When writing out SDF file from DC, you can specify whether the Load Delay should be included in the IOPATH delay or INTERCONNECT by using the appropriate options with the write_timing command. Also, if the resistance value in the wire-load model selected is 0, the Connect Delay component will be 0. Shown below is a Synopsys library description of the wire-load model.

```
wire_load("10x10") {
  resistance : 0 ;
  capacitance : 1 ;
  area : 0 ;
  slope : 0.311 ;
  fanout_length(1,0.53) ;
}
```

Further, if the tree_type in the operating conditions selected is best_case_tree, the wire resistance is taken as 0 and the Connect Delay is 0. Shown below is a Synopsys library description of the operating conditions.

```
operating_conditions(BCCOM) {
  process : 0.6 ;
  temperature : 0 ;
  voltage : 5.25 ;
  tree_type : "best_case_tree" ;
```

If no operating conditions are specified, the default tree_type assumed is balanced_tree.

**Scenario 11**

You wish to add a user-defined prefix to all the cells in the netlist. But the define_name_rules command with the -prefix option does not do this. Why?

**Solution**

The -prefix option is only used when change_names needs to create a completely new name to insure the name is unique. It is used with the max_length option for situations where in order to generate unique object names, the original name has to be deleted and names of the form <prefix><index> are generated until a unique name is found. The <prefix> is defined using the -prefix option.

**Scenario 12**

When attempting to write out a design in db format, you get the following error message:

```
write top
Error: 'top' doesn't specify a unique design
Please use complete specification:full_file_name:design_name
```

**Solution**

You can read more than one design with the same name into dc_shell. In this case, it is likely that there are two designs with the file name top.db in dc_shell (memory). To confirm this list, the designs in dc_shell use the following command:

```
list -designs
```

To avoid ambiguity, specify the file name when writing the design. To resolve this conflict, precede the file name with its full pathname. In the case of a design name conflict, use the file name to write the design as follows:

```
write -f db top.db:top
```

## Recommended Further Readings

1. Synopsys EDIF Interface User Guide

2. Synopsys "Impact" Quarterly newsletter Q&A section.

CHAPTER 10

# *Design Re-use Using DesignWare*

This chapter describes the Synopsys DesignWare concept introduced in Chapter 1. DesignWare allows one to use pre-existing components to easily implement designs. This chapter also discusses the mechanism for inferring complex cells using DesignWare. The steps involved in building your own DesignWare library are outlined. Finally, classic scenarios involving DesignWare are described and solutions provided.

DC can only infer combinational DesignWare components. Sequential DesignWare components must be instantiated in order to be used in designs. However, the Synopsys *Behavioral Compiler* (refer to Chapter 11) supports RAM inferencing and inference of pipelined components. To support *Behavioral Compiler* capabilities, enhancements have been made to DesignWare to allow modelling of cycle by cycle behavior of sequential DesignWare components in the DesignWare synthetic library.

## 10.1 DesignWare Libraries

The DesignWare library is a library of reusable DesignWare components, such as adders, subtracters, multipliers, counters, and FIFOs. Reusable components play a significant role in speeding up the design cycle. DW03_UPDN_CTR(up down counter), DW03_FIFO_S_DF (FIFO), DW03_SHFTREG (shift register), DW01_decode (decoder), and DW03_PIPE_REG (pipeline register) are a few examples of components from the Synopsys DesignWare libraries. Further information about these and other components are available in the DesignWare component databooks.

Synopsys provides the simulation models for DesignWare components (VHDL models for all components and Verilog models for some), making the task of simulation simpler and several times faster. In addition to the components provided in the DesignWare libraries, it is possible to develop and re-use design modules by building a proprietary DesignWare library.

## 10.2 Inferring Complex Cells

In this section, we discuss certain common scenarios with regard to inferring complex cells. This helps one to understand the DesignWare mechanism and how it can be used to infer complex cells which are difficult to infer via synthesis.

**Scenario 1**

You have a certain cell available in your technology library, but DC does not infer this cell during synthesis.

**Solution**

The crude, yet working solution is to instantiate the cell and `dont_touch` the cell during synthesis.

If DC does not infer a certain cell from the technology library, this is most likely due to one of the following reasons: the Synopsys library does not have a function description for that cell, and hence, DC considers the cell as a black box, or the cell has a function attribute, but DC is incapable of mapping to this particular cell.

**Scenario 2**

You have several implementations of a certain datapath module (say an adder.) You wish to use a different implementation (the most appropriate with regard to speed and area) in each instance of an adder in the design. For example, you wish to use a carry select adder in one block, but a ripple carry implementation of the same adder in another.

**Solution**

Instantiate the appropriate implementation and `dont_touch` the instance if the implementation is an instantiation of a library cell. If you have more than one implementation available then you have to manually select the most appropriate implementation to instantiate which is cumbersome.

**Scenario 3**

You have designed a module for a particular block in a design, and wish to re-use this exact same module in another block. However, you require a larger bus width parameter for this module. Further, you wish to use the most optimal implementation based on the bus-width parameter. A typical example, would again be an adder module. For example, in one block you wish the module to be a ripple carry adder of 8 bits, while in another block you wish to use an adder of 16 bits.

**Solution**

For each of the scenarios discussed above, DesignWare provides an effective solution as described in section 10.3.

### 10.2.1 Mapping to DesignWare Components

In DC, arithmetic operations like +, -, *, <, >, <=, and >= are implemented using DesignWare components. When reading in your Verilog or VHDL code into DC using (V)HDL Compiler, these predefined operations are automatically mapped to DesignWare implementations of these operations. By default, the Synopsys DesignWare library, standard.sldb, is always used during the synthesis process. There is no way to turn off this library except by placing a dont_use attribute on the synthetic modules available in this library. There are several advantages to have these arithmetic operations directly map to DesignWare components. First, this ensures appropriate sharing of resources. Second, selection of the implementation (ripple adder vs. carry look-ahead adder) will be based on design constraints. Further, mapping to DesignWare components often results in shorter compile run times. Most importantly, DesignWare implies optimization at a higher level of abstraction.

A DesignWare module once built for a particular technology library, (also for different parameters) is cached in the location pointed to by the cache_write variable. This means that once the components have been built and cached, successive compile runs will be a lot faster because these parts do not need to be built again. The Synopsys variable cache_read points to the location of the Synopsys cache from which it can read pre-cached DesignWare components. The Synopsys cache can be shared among the designers in a group by setting the cache_read and cache_write variables appropriately. Otherwise, a separate cache (default being the users home directory) will be created with the same components for each user.

Operations like =, /=, and logical operations are not mapped to DesignWare components and hence implemented using random logic. Further, they are not shared as resources and cannot be saved in a cache. They are built every time you compile a design. However, it is possible to perform DesignWare encapsulation for random logic using function/procedure calls.

Example 10.1 shows the mechanism by which the predefined arithmetic operation, "*" in VHDL is mapped to DesignWare components. The "*" operation in the VHDL code for data type *unsigned* operands is mapped to the corresponding function in the std_logic_arith package. This function in turn calls the function mult which is mapped to the DesignWare synthetic operator MULT_UNS_OP.

### Example 10.1 Arithmetic Operation Mapped to DesignWare Component

```
function "*"(L: UNSIGNED; R: UNSIGNED) return UNSIGNED is
-- pragma label_applies_to mult
-- synopsys subpgm_id 295
   begin
      return mult(CONV_UNSIGNED(L, L'length),
         CONV_UNSIGNED(R, R'length));
-- pragma label mult
   end;
function mult(A,B: UNSIGNED) return UNSIGNED is
    variable BA: UNSIGNED((A'length+B'length-1) downto 0);
    variable PA: UNSIGNED((A'length+B'length-1) downto 0);
    constant one : UNSIGNED(1 downto 0) := "01";
    -- pragma map_to_operator MULT_UNS_OP
    -- pragma type_function MULT_UNSIGNED_ARG
    -- pragma return_port_name Z
    begin
-- FOR SIMULATION PURPOSES ONLY
    end;
```

If you do not use the VHDL packages provided by Synopsys and create your own functions for operations like "*", they will not be mapped to DesignWare components but instead inferred from random logic. This could also happen when you accidently forget to include, for example, the std_logic_arith package. For example, you are reading in the VHDL code segment in example 10.2 and have not included the Synopsys packages which defines the "<" operation for the data types of a and b, which is std_logic_vector. Note that this is not applicable for Verilog designs.

The code in Example 10.2 can be read into DC and synthesized without any errors, though the < operation will be built using random logic. This is because, for scalar data types, the predefined relational operators are implicitly defined.

**Example 10.2 Arithmetic Operation Built from Random Logic**

```
process (a, b)
begin
  if (a < b) then
     out1 <= something1;
  else
     out1 <= something2;
  end if;
end process;
```

### 10.2.2 Inferring cells using procedure calls

A common complaint in the design community is a lack of control over the cells inferred during synthesis. A combination of procedure (function) calls and DC directives such as the map_to_entity directive can give the user a fair amount of control over the cells inferred. This helps tackle the issue discussed in Scenario 1 in section 10.1.

Example 10.3 shows how one can achieve the required functionality via a procedure call. The procedure latch_proc is defined in package my_pack. This package is analyzed into the work library. When a procedure call to latch_proc is made in the VHDL code, the map_to_entity directive in the procedure latch_proc ensures that the LD1 latch is inferred from the technology library. Thus, the procedure implies a certain implementation. This ensures that the required functionality is achieved using the desired library cell. Expanding on this approach, one can develop a library of procedures and functions and easily force the inference of certain technology specific implementations during synthesis.

## Example 10.3 Component Implication Directives on a Procedure

```
-- Component Implication Directives :
library IEEE;
use IEEE.std_logic_1164.all;
package my_pack is
  procedure latch_proc(D: in std_logic;
G: in std_logic;
signal Q: out std_logic;
signal QN: out std_logic);
end my_pack;
package body my_pack is
  procedure latch_proc (D: in std_logic;
 G: in std_logic;
 signal Q: out std_logic;
 signal QN: out std_logic) is
-- pragma map_to_entity LD1
-- The procedure body is used for simulation purposes
-- only
  begin
if (G = '1') then
  Q <= D;
  QN <= not D;
end if;
  end latch_proc;
end my_pack;
-- Entity which uses the above defined procedure
use work.my_pack.all;
library IEEE;
use IEEE.std_logic_1164.all;
entity test is
  port (data, enable: in std_logic; q, qbar: out std_logic);
end test;
architecture behv of test is
begin
  process(data, enable)
  begin
-- Instantiating components in process statements due to the
-- map_to_entity pragma. This would not be allowed normally
-- in VHDL
latch_proc (data, enable, q, qbar);
  end process;
end behv;
```

# 10.3 Creating Your Own DesignWare Library

DesignWare, in addition to its other advantages, provides an effective mechanism to address Scenarios 2 and 3 described in section 10.1. For example, you have several implementations of a module, and wish that the synthesis tool infer the most optimal implementation. This can be achieved using DesignWare libraries. Alternately, if you rather make the choice yourself than let the tool select an implementation, this can be achieved by placing a dont_use attribute on all the undesired implementations or by using the set_implementation command.

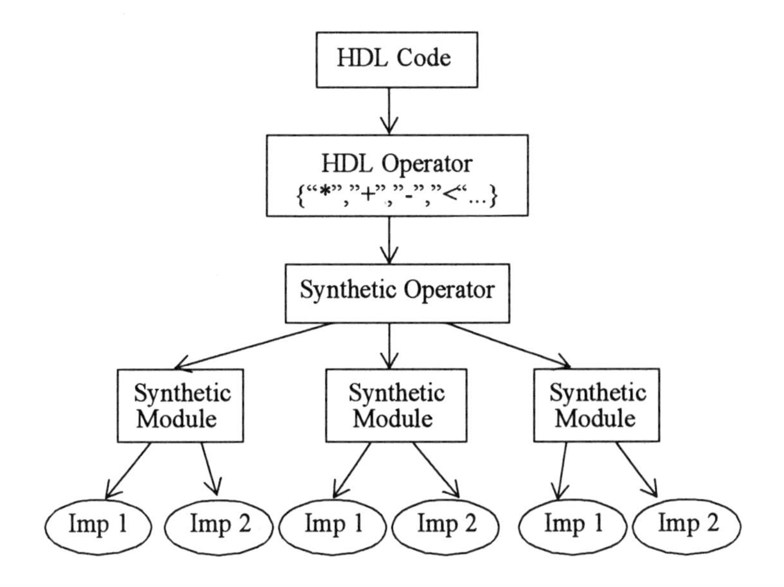


**Figure 10.1 Link from HDL to DesignWare Libraries**

Figure 10.1 shows the DesignWare approach to achieving specific implementations. In the HDL code for a procedure similar to the latch_proc procedure, if one were to use a map_to_operator directive instead of a map_to_entity directive, the tool would infer the required synthetic operator on reading in the HDL to DC. One must then have a DesignWare library which provides the link between the operator and one or more synthetic modules. Each operator can be linked to multiple synthetic modules and each module can have multiple implementations. Based on the constraints specified, the DC ensures that the most optimal implementation is inferred.

A synthetic operator is essentially an intermediate stage between reading the HDL source and before a compile step. During the compile step, the inferred synthetic operators are synthesized to one or more technology library cells. To find the different synthetic operators and DesignWare modules inferred, read in a source VHDL/Verilog and execute the report_reference command. The tool should list the different operators, if any. For example, the ADD_UNS_OP is an operator from the DW01, built-in Synopsys DesignWare library.

Example 10.4 shows the implementations for MULT_UNS_OP. Information about modules, operators and the implementations available can be obtained by executing the report_synlib command. The synthetic operator MULT_UNS_OP is defined in the standard.sldb synthetic library. MULT_UNS_OP may be mapped to multiple synthetic modules which may have more than one implementation. The synthetic module DW02_mult has among other bindings, one binding b3 to MULT_UNS_OP.

## Example 10.4 Implementations for MULT_UNS_OP

```
DW02_mult
        b1    bound_operator: MULT_TC_OP
              Pin Associations (module, oper):
                A, A
                B, B
                TC,"1"
                PRODUCT, Z
        b2    bound_operator: MULT_TC_OP
              Pin Associations (module, oper):
                A, B
                B, A
                TC,"1"
                PRODUCT, Z
        b3    bound_operator: MULT_UNS_OP
              Pin Associations (module, oper):
                A, A
                B, B
                TC,"0"
                PRODUCT, Z
        b4    bound_operator: MULT_UNS_OP
              Pin Associations (module, oper):
                A, B
                B, A
                TC,"0"
                PRODUCT, Z
```

### 10.3.1 Requirements for Building a DesignWare Library

Consider the case of an adder module, with four possible implementations – behavioral, fast, faster, and a fastest implementation.

1. A VHDL/Verilog file declaring the add function along with the map_to_operator directive.

**VHDL Code**

```
package my_pack is
PROCEDURE add (    a, b:      IN std_logic_vector;
            SIGNAL sum:     OUT std_logic_vector;
                ci:     IN std_logic;
            SIGNAL co:     OUT std_logic );
end my_pack;
package body my_pack is
PROCEDURE add ( a, b: IN std_logic_vector; SIGNAL sum: OUT std_logic_vector;  ci:
                                    IN std_logic ; SIGNAL co: OUT std_logic ) is
     -- pragma map_to_operator ADD1_OP
begin
---- include code for add procedure
end ;
end my_pack;
```

2. A VHDL/Verilog file describing the adder module with the different possible architectures described. The adder module (and other such modules) is referred to as a synthetic module.

**VHDL Code**

```
entity add1 is
    port ( A : in std_logic_vector (width -1 downto 0);
           B : in std_logic_vector (width -1 downto 0);
          cin : in std_logic ;
           sum: out std_logic_vector (width -1 downto 0);
           co : out std_logic );
end add1 ;
architecture behavior of add1 is
begin
--- include behavioral description of add1 module
end behavior;
architecture fast of add1 is
begin
--- include fast implementation of add1 module
end fast;
architecture faster of add1 is
begin
--- include faster implementation of add1 module
end faster;
architecture fastest of add1 is
begin
--- include fastest implementation of add1 module
end fastest;
```

The concept of architectures available in VHDL does not exist in Verilog. Instead, each implementation of a synthetic module is a separate Verilog module. Hence, Verilog DesignWare implementations must be of the form:

```
synthetic_module_name__implementation_name
```

Parameters used in Verilog must correspond to synthetic module parameters. The synthetic module names must match the module name in Verilog (as in VHDL), and the hdl_parameter attribute of the synthetic module parameter must be set to TRUE.

Example 10.5 shows the template for a synthetic module adder with four implementations behavior, fast, faster and fastest in Verilog.

## Example 10.5 Template for a Synthetic Adder

### Verilog Code

```
module adder1__behavior (a, b, cin, sum, co);
input a, b, cin;
output sum, co;
// implementation 1 description
endmodule
module adder1__fast (a, b, cin, sum, co);
input a, b, cin;
output sum, co;
// implementation 2 description
endmodule
module adder1__faster (a, b, cin, sum, co);
input a, b, cin;
output sum, co;
// implementation 3 description
endmodule
module adder1__fastest (a, b, cin, sum, co);
input a, b, cin;
output sum, co;
// implementation 4 description
endmodule
```

3. The .sl or synthetic library file lists the operators (called synthetic operators), the pin configuration of the adder module, the parameters and the *binding* between the operator, the modules and the different possible implementations. This .sl file must then be compiled using the read_lib (followed by write_lib) command. This should generate the synthetic_library (the .sldb file). All the modules defined in the synthetic library are called synthetic modules. The synthetic_library, by default, is set to standard.sldb. There is no way to turn off the standard.sldb during compile. In other words, setting the variable synthetic_library = {} will not prevent the DC from inferring parts from the standard.sldb synthetic library. But the dont_use command can be used to disable specific implementations.

Example 10.6 shows a sample synthetic library (.sl file) with one synthetic operator ADD1_OP, and one synthetic module add1.

## Example 10.6 Sample Synthetic Library

```
library ("mylib.sldb"){
operator("ADD1_OP"){
pin(a){
        direction : input;
    }
pin(b){
        direction : input;
    }
pin(ci){
        direction : input;
    }
pin(sum){
        direction : output;
    }
}
module(add1){
    library : "personalware";
    parameter(width){
    formula : "width"
    hdl_parameter : TRUE;
     }
    implementation("behavior"){
     }
    implementation("fast"){
     }
    implementation("faster"){
     }
    implementation("fastest"){
     }
pin(a) {
    direction : input;
    bit_width : "width";
}
    pin(b) {
    direction : input;
    bit_width : "width";
}
pin(ci) {
    direction : input;
    bit_width : "1";
}
```

```
pin(co) {
      direction : output;
     bit_width : "1";
}
pin(sum) {
      direction : output;
      bit_width : "width";
}
binding(b1){
        bound_operator : "ADD1__OP";
           pin_association(a)  { oper_pin : a; }
           pin_association(b)  { oper_pin : b; }
           pin_association(ci) { oper_pin : ci; }
           pin_association(co) { oper_pin : co; }
           pin_association(sum){oper_pin : sum; }
}
}
```

You may either allow DC to evaluate and select the best implementation or use the set_implementation command to select your implementation. Further, there are attributes available in Designware like priority and legality which allow you to specify the priority among different available implementations (rather than permitting DC to evaluate them) and the parameters for which that implementation is a legal (allowed).

Example 10.7 describes an .sl file which shows the use of the "legality" attribute. This .sl file is also defining a new user-defined implementation for a synthetic module "DW01_decode" which is only a legal implementation to be considered by DC when "width = 8".

## Example 10.7 Legality Attribute

```
library ( ext_imp.sldb ) {
   implementation(my_decode,DW01_decode) {
     technology : class.db;
     parameter (legal) {
       formula : "width == 8";
     }
   }
}
```

## 10.4 Classic Scenarios

**Scenario 1**

DC issues the following warning when executing the replace_synthetic command.

W LINK-9 Unable to resolve reference to synthetic module '%s' in '%s'.

**Solution**

This warning occurs when the DC is unable to find a particular synthetic module in the synthetic library. If you have created your own synthetic library, execute the following command as a verification step.

```
report_synlib <name of synthetic library>
```

Verify that the synthetic module specified in the warning message exists in the synthetic library.

**Scenario 2**

DC issues the following error message.

Error: Can't find implementation fast_adder in library personalware.

**Solution**

This error occurs when the VHDL/Verilog file which describes the fast_adder implementation has not been analyzed into the personalware library or when the synthetic library source file (that is, the .sl file) does not have the fast_adder implementation declared for the particular module.

**Scenario 3**

You have instantiated the DesignWare component DW03_updn_ctr in your code. On compile, DC gives the following error:

Cannot find a valid implementation for processor DW03_updn_ctr (SYNH-14)

**Solution**

This could be due to the fact that the user does not have a license for SynLib-Seq. This part can however, be evaluated and timing and area reports obtained from it by setting the variable

```
synlib_disable_limited_licenses = false
```

This will allow the component to be instantiated and compiled, but no netlist can be written out, nor will the schematic representing the part be accessible.

**Scenario 4**

You have a design with complex cells instantiated. These cells are black boxes for DC since the Library Compiler does not support its functionality. There are different drive and timing versions of these cells in the target technology library. How does one ensure that DC resizes these cells when optimizing for timing or area?

**Solution**

DC supports an attribute called user_function_class for cells that cannot be functionally modeled in Synopsys. Black box cells with the same user_function_class attribute and the same number of pins are treated functionally equivalent in DC. DC will then be able to resize these cells provided there are timing arcs to the output pins of the cells provided in the target technology library. This attribute can be specified in the library source code itself or by using the set_attribute command in DC.

**Scenario 5**

You are synthesizing your design and are using DesignWare library provided by an external vendor. Synthesis infers a combinational synthetic module with a particular implementation. But your timing constraints are not met. There are faster implementations available in the library for this synthetic module. How can I infer a faster implementation?

**Solution**

By default, DC works on the worst violator in each path group. Verify if the datapath which includes this synthetic module is the worst violator. If not, then use the group_path command to create a separate path group for this path with a higher weight. This causes the violated path to be included in the synthesis cost function and could result in a faster implementation from being inferred. You can prevent DC from using a particular implementation, by placing a dont_use attribute on the other implementations.

**Scenario 6**

You are creating a synthetic module implementation which instantiates cells from a ASIC vendor library libA. How can you ensure that this particular implementation of the synthetic module is chosen only when the user's target_library variable is set to libA.

**Solution**

Use the technology attribute in the synthetic library to define the target technology of the implementation. If the technology attribute is set to libA, DC will select that particular implementation only if the library listed first in target_library matches the technology attribute. Shown below is an example section of the synthetic library:

```
implementation (my_imp) {
  technology : libA.db;
}
```

**Scenario 7**

You want to add two 8-bit numbers A and B using a 8-bit DesignWare adder such that carry out bit of the DesignWare adder is used to give a 9 -bit result.

**Solution**

The following code template can be used to generate this:

```
module test(a, b, sum);
input [7:0] a;
input [7:0] b;
output [8:0] sum;
wire [8:0] atemp, btemp;
assign atemp = {0, a};
/* Make a & b 9-bit numbers by concatenating 0 */
assign btemp = {0, b}; /* to the most significant bit */
assign sum = atemp + btemp + 1;
endmodule
```

**Scenario 8**

You are creating a DesignWare library and licensing out the parts to other customers. You want to encrypt the VHDL/Verilog models and then send them out to the customers. Is this possible?

**Solution**

Synopsys has a utility named synenc that allows users to encrypt VHDL or Verilog DesignWare models. The encrypted VHDL DesignWare models can be analyzed by the VHDL Compiler or the vhdlan utility and then elaborated in the Synopsys synthesis or simulation tools. The encrypted Verilog DesignWare models can be analyzed by the HDL Compiler and then elaborated in the Synopsys synthesis tool. The synenc utility requires a separate license and a DesignWare-Developer license in order to be used.

**Scenario 9**

You have read in your source HDL followed by compile. An add function in the source HDL code mapped to an adder from the Synopsys DesignWare library. The current implementation is a ripple adder (rpl), but you require a carry-lookahead adder (cla) to be inferred. Does one have to start with reading in the source HDL and specifying constraints to change the implementation to a cla adder.

**Solution**

Use the set_implementation command to specify the cla implementation. You do not have to read in the source HDL to set the implementation. Execute the report_resources command. This should report the adders, muxes, and other resources used in the design. Once you know the instance name and the required implementation, use the set_implementation command as follows:

```
set_implementation cla U1
```

U1 is the instance name of the adder module and cla the name of the desired implementation. Then, re-compile the design.

**Scenario 10**

You wish to attach an attribute to a Designware component. How do you go about doing this?

**Solution**

Below is an example, where the attribute is_ram is specified on an entity through an embedded dc_shell script. The attribute can also be specified from dc_shell.

```
 entity COUNT_SEQ_VHDL is
  port(DATA, CLK: in BIT;
      RESET, READ: in BOOLEAN;
      COUNT: buffer INTEGER range 0 to 8;
      IS_LEGAL: out BOOLEAN;
      COUNT_READY: out BOOLEAN);
end;
architecture BEHAVIOR of COUNT_SEQ_VHDL is
-- pragma dc_script_begin
-- set_attribute current_design is_ram true -type boolean
-- pragma dc_script_end
begin
  process
    variable SEEN_ZERO, SEEN_TRAILING: BOOLEAN;
    variable BITS_SEEN: INTEGER range 0 to 7;
  begin
```

```
--- ARCHITECTURE DESCRIPTION
end process;
end BEHAVIOR;
```

**Scenario 11**

Your VHDL design instantiates a component "SUB" from the technology library. The component declaration of the component "SUB" is in a package "comps" which is made visible by the use clause. I get this error during read.

```
I1: SUB port map (a, b, sum);
    ^
**Error: /home/vhdl_code/test.vhd  line 26
    Identifier is not visible because it is ambiguous (homographs were introduced by USE
clauses). (VSS-574)
```

**Solution**

"SUB" is an enumeration literal defined in type "CHARACTER" in the package "STANDARD" which is implicitly visible in all VHDL designs. Since this definition as well as the component declaration for "SUB" are both visible in the code, a homograph is reported. You can explicitly avoid this ambiguity as follows

```
SUB_label : WORK.COMPS.SUB generic map (width => 1) port map (a, b, sum);
```

**Scenario 12**

You are using Designware Developer to create a synthetic operator "EQ_UNS_OP"; synthetic module and implementations of the "==" operation. How can one automatically infer this Designware component for the "==" operation in the Verilog code.

**Solution**

The "==" operation in Verilog is not mapped to a DesignWare component. One way to map the "==" operation to your Designware component is by using the "map_to_operator" pragma in a function call as shown in the example below:

**Verilog Code**

```
module eq (x, y, z);
input x, y;
output z;
 reg z;

function equal;
```

```
// synopsys map_to_operator EQ_UNS_OP
// synopsys return_port_name Z
input A;
input B;
begin
 equal = (A == B);
end
endfunction

always @ (x or y)
begin : b1

       z = equal(x,y);
end

endmodule
```

## Recommended Further Readings

1. DesignWare User Guide
2. DesignWare Components Databook
3. DesignWare Developer Guide
4. Synopsys "Impact" quarterly newsletter. Q&A section.

# CHAPTER 11 *Behavioral Synthesis - An Introduction*

Behavioral synthesis has been widely touted as the next major step in design automation after logic synthesis. Several behavioral synthesis tools are commercially available. However, a large percentage of logic designers still follow schematic capture based design methodology. This clearly raises some extremely pertinent issues. Are behavioral synthesis tools ahead of their times? Are these tools easy to use? How do these tools fit into the ASIC design flow? Is there a growing user base for these tools? And most importantly, do *you* need behavioral synthesis?

This chapter provides an introduction to behavioral synthesis. First, the motivation for behavioral synthesis is discussed with logic synthesis in perspective. Then, the behavioral synthesis based design flow is described. This is followed by a simple example to illustrate the flow. The bubble sort, sorting algorithm is the example discussed. The behavioral synthesis tool used to illustrate this example is the *Behavioral Compiler* (BC) from Synopsys. Mistral 1&2 (from Mentor Graphics) and ViewSchedule (from Viewlogic) are two other commercially available behavioral synthesis tools from leading EDA vendors.

## 11.1 Logic Synthesis

In comparison to schematic capture tools, logic synthesis has helped speed up the design cycle time and improve productivity. However, there are some inherent limitations to logic synthesis. HDL descriptions which can be synthesized using commercial logic synthesis tools are often referred to as RTL (register-transfer level) code or synthesizable code. RTL descriptions imply clocks and registers, distinctly binding operations to specific clock cycles. This implies that the different architectural choices available to implement the design are limited by the HDL description. In other words, when coding for logic synthesis, a certain architecture is pre-defined or implied, thereby, limiting the evaluation of alternatives. Hence, to explore multiple architectures for a design one would have to re-write the RTL code. This is often an extremely complicated and time-consuming task. Logic synthesis tools in general, optimize a design to meet timing and area constraints, not to mention synthesis tools that optimize for power consumption. Further, logic synthesis performs optimization of combinational logic between registers and does not optimize the usage of the registers themselves.

RTL synthesis and its limitations can be illustrated using a simple multplier as shown in Example 11.1. The multiplication operation (*) has been explicitly assigned to a specific clock cycle. If the delay through the multiplier is greater than the clock period, there will be a timing violation. The logic synthesis tool will try to select the fastest possible implementation of a multiplier to meet timing. On the other hand, if the delay through the multiplier is much less than the clock cycle, then logic synthesis does not take advantage of this to assign additional operations to this clock cycle.

## Example 11.1 Multiplication Operation Assigned to a Clock Cycle

**Verilog Code**

```
always @(posedge clock)
begin
outp <= a * b;
end
```

**VHDL Code**

```
process
begin
wait until clock = '1' and clock'event;
outp <= a * b;
end process;
```

Behavioral synthesis could address this problem by either inferring pipelined multipliers or by automatically making the multiplication operation a multicycle path. Similarly, behavioral synthesis can *chain* operations if the delay of the multiplication operations is less than the clock period and thereby reduce latency and register costs. In other words, behavioral synthesis can move operations across clock cycles based on timing requirements, provided no control or data dependencies are violated.

Logic synthesis of one dimensional and multi-dimensional arrays in VHDL or Verilog results in *x times y* registers for an array with x and y dimensions. Further, to write to an element of the array requires decoding logic and to read an element from the array requires multiplexing logic. Behavioral synthesis tools are capable of directly mapping arrays in VHDL/Verilog to RAM cells provided they are available in the target technology library.

# 11.2 Behavioral Synthesis Concepts

In this section, we briefly introduce certain basic concepts which help understand behavioral synthesis better.

## 11.2.1 Latency and Throughput

Throughput is defined as the rate at which a system can sample new data for processing. Latency on the other hand is the total number of clock cycles required to execute all operations in a single loop iteration. The fundamental objective of behavioral synthesis is to assist the designer in arriving at an optimal architectural description from several different functionally equivalent implementations that meet the latency and clock period requirements. In addition, to area and timing behavioral synthesis implies optimization in another dimension, namely, latency.

## 11.2.2 Register Optimization and Memory Inferencing

The behavioral synthesis is capable of both register optimization and memory inferencing. It can evaluate the life time of a variable declared in the HDL and share registers for variables which have different life times. This implies that the same functionality can be inferred using lesser number of registers. This is called register optimization. Memory inferencing means that arrays in the HDL code can automatically infer memory hard macros and the associated control logic. In RTL synthesis on the other hand, one has to explicitly instantiate the memory macros and describe the intricate control logic.

## 11.2.3 Chaining of Operations

Behavioral synthesis *chains* operations within a cycle based on the clock period. The following VHDL design example illustrates the concept of chaining.

### Example 11.2 Chaining of Operations

```
t1 := inp1 * inp2;
t2 := func(t1, inp3); -- assume func is mapped to a combinational DW module
sig1 <= t2;
wait until clock = '1' and clock'event;
```

Consider Example 11.2 when synthesized using BC. First, the two functions * and func are mapped to DesignWare modules. Then, BC estimates the timing requirements of these modules. Based on the clock period and other scheduling constraints, BC ascertains whether it can perform (chain) these operations in one clock cycle or not. Changing the clock period affects the chaining of operations. Increasing the clock period aids chaining, but does not encourage sharing of resources. Similarly, reducing the clock period does not aid chaining but encourages sharing of resources, when possible. Thus, behavioral synthesis allows one to investigate different clock periods and select the implementation that results in the most appropriate balance between datapath operations and register cost.

### 11.2.4 Pipelining

Pipelining is the process of adding registers to meet clock timirng requirements at the expense of a slower latency. To understand pipelining better, consider a simple design which involves a block of combinational logic between two registers as shown in Figure 11.1

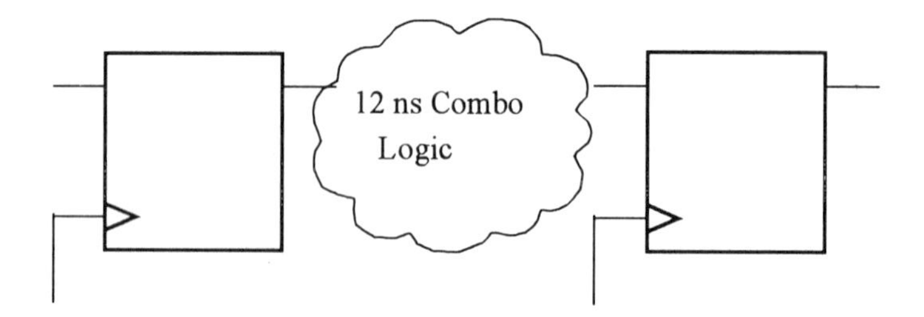


**Figure 11.1 Combinational Logic Between Registers**

Let us assume that the combinational logic has a total delay of 12 ns, a clock period of 10 ns and a setup time requirement of 0 ns. In general, logic synthesis optimizes combinational logic between registers to meet the required clock period. Hence, in this example, logic synthesis will attempt to reduce the delay in the combinational block to less than 10 ns. However, if logic synthesis is unable to do so, the synthesized design will show setup timing violations.

Behavioral Synthesis assigns operations to a clock cycle based on the delay of the operations and the clock period. If the delay through a single operation exceeds the clock period then the Behavioral synthesis tool will automatically make it a multi-cycle operation and not chain any additional operations with the multicycle operation. Alternatively, one can inferring pipelined components from the DW library. For example, Synopsys provided DW libraries have pipelined multipliers with different number of pipelined stages. These can be inferred when the delay though the multiply operation exceeds the clock period and one does not desire a multicycle operation. One can also build a proprietary DW library (as discussed in chapter 10) with piplined DW components. In the above example, the Behavioral Synthesis tool automatically add registers between operations if the sum of the delay s through them exceeds the clock period.

## 11.3 Synopsys Behavioral Compiler

In this section, we briefly describe the features of the Synopsys Behavioral Compiler (BC). A clear understanding of behavioral synthesis concepts described in section 11.2 should assist the reader in understanding BC. BC performs *automatic scheduling of operations* (when the corresponding operators are mapped to DesignWare components) such as addition, multiplication, memory reads and memory writes to clock cycles. Further, it *pipelines operations* provided a pragma is specified in the HDL. BC can automatically schedule the READ and WRITE cycles of the RAMs. This is possible provided, information on the number of cycles required for read/write, number of data ports in the RAM and cycle to cycle pin connection are available. In Synopsys, this cycle to cycle information of the RAMs is provided in the DesignWare synthetic library. In general, DC only supports inference of combinational designware components while BC can infer both combinational and sequential DesigWare components.

Timing requirements are met by *chaining operations,* through multi-cycle operations, and by *adding or sharing resources*. BC also ensures *register optimization, memory and register inferencing*. Most importantly, behavioral synthesis implies design capture at a *higher level of abstraction* and ensures *faster simulation.*

BC also *automatically generates the control FSM* required to control the datapath logic. Thus, one does not have to create any complex state diagrams. Based on the higher-level abstraction code, it makes *automatic scheduling* decisions while tool variables or switches control pipelining and chaining. Also, BC provides the capability to generate implementations with minimum latency or an implementation which meets all scheduling constraints with a minimum area. In short, BC can be used to generate a *host of design alternatives* and the user can then select the implementation depending on the target application.

## 11.4 Behavioral Synthesis Design Flow

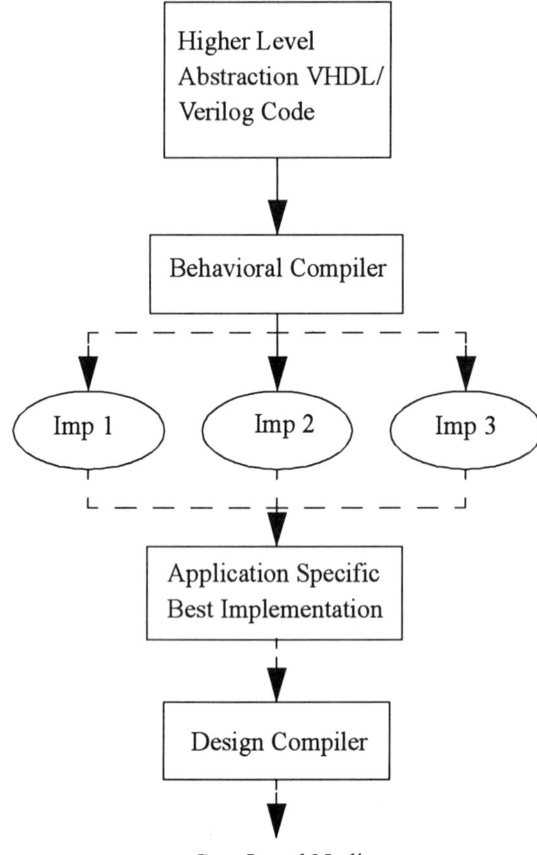


**Figure 11.2 Design Flow from Algorithmic Description To Gates Using Behavioral Compiler**

The input to BC is a high-level algorithmic description (behavioral), while the output is an optimized RTL description which is then optimized to gates using DC as shown in Figure 11.2.

BC accepts both VHDLand Verilog. The behavioral HDL description is converted to an intermediate database, the Synopsys db format. In effect, an HDL written at higher level of abstraction is converted to multiple RTL implementations using behavioral synthesis. All the information about constraints and multi-cycle operations are embedded in this design database. Then, logic synthesis is used to map the database to gates. Hence, it is extremely useful for the behavioral synthesis tool to be tightly integrated with a logic synthesis tool.

## 11.5 Example Using Behavioral Compiler

Behavioral synthesis begins with high-level algorithmic descriptions in VHDL or Verilog. In other words, behavioral synthesis implies design capture at a higher-level of abstraction than RTL. Thus, from the perspective of a designer, behavioral synthesis differs from present-day logic synthesis in the coding style.

For behavioral synthesis, functionality is described with regard to occurrence of operations with no notion of the actual implementation and without binding operations to specific clock cycles. Architectural implementation decisions are made by the behavioral synthesis tool based on cycle period, latency and throughput goals. The behavioral synthesis tools does however, maintain the implicit dependency of the operations in the HDL code.

Example 11.3 shows the sorting algorithm in C programming language and the corresponding behavioral (algorithmic) description in VHDL/Verilog for a simple bubble sort algorithm. Example 11.4 shows the scripts required to synthesis the design to gates. Bubble sort is an exchange sort that involves repeated comparison and, if necessary, the exchange of adjacent elements. Sorting is the process of arranging a set of similar information into an increasing or decreasing order.

The VHDL entity in Example 11.3 has an input which is an array of integers to be sorted. First, this array is loaded into an internally inferred RAM. Then, this array is sorted and later saved back in the same RAM. Finally, the sorted list is shifted out. The behavioral VHDL description illustrates RAM inferencing, control over rolling and unrolling of *for* loops (in logic synthesis all *for* loops are unrolled or in other words, DC generates hardware for each iteration of the *for* loop in parallel), synchronous reset inferencing and in general, VHDL coding at a much higher level of abstraction (algorithm). The attributes used in this example are specific to BC.

### Example 11.3 Bubble Sort Algorithm

#### C Language Code

```
"/* The Bubble Sort */
void bubble(char *item, int count)
{
  register int a, b;
  register char t;
```

```
  for(a=1; a < count; ++a)
for(b = count - 1; b >= a; --b){
  if(item[b-1] > item[b]) {
  /* exchange elements */
  t = item[b-1];
  item[b-1] = item[b];
  item[b] = t;
  }
}
 }
```

**VHDL Code**

```
 1-- Implementing the bubble sort algorithm in VHDL
 2-- Using a RAM for internal 70 bit register
 3
 4 library IEEE;
 5 use IEEE.std_logic_arith.all;
 6
 7 package my_pack is
 8
 9 subtype int_limit is unsigned(6 downto 0);
10 type data_type is array(NATURAL range <>) of int_limit;
11
12 end my_pack;
13
14
15 library IEEE;
16 use IEEE.std_logic_1164.all;
17 use IEEE.std_logic_arith.all;
18 use work.my_pack.all;
19 library dware;
20 use dware.behavioral.all;
21 library synopsys;
22 use synopsys.attributes.all;
23
24 entity bubble_sort is
25 -- Number of elements in the array is parametrizable
26   generic (N: NATURAL := 10);
27   port (unsorted_data : in unsigned(6 downto 0);
```

```
28       clk, reset: in std_logic;
29       sorted_data: out unsigned(6 downto 0));
30  end bubble_sort;
31
32
33  architecture behv of bubble_sort is
34
35  begin
36
37   TOP_LOOP: process
38   variable index2: natural range 0 to 10;
39   variable temp1, temp2: unsigned(6 downto 0);
40   constant RAM_A : resource := 0;
41   attribute variables of RAM_A : constant is "unsorted_reg";
42   attribute map_to_module of RAM_A : constant is "DW03_ram1_s_d";
43   variable unsorted_reg: data_type(1 to N);
44   attribute dont_unroll of loop1 : label is true;
45   attribute dont_unroll of ram_load : label is true;
46   attribute dont_unroll of ram_read : label is true;
47   attribute dont_unroll of ram_init : label is true;
48   begin
49     reset_loop: loop
50
51-- Initialize all internal variables and output ports
52
53  sorted_data <= "0000000";
54  ram_init: for ram_index in 1 to N loop
55   wait until clk = '1' and clk'event;
56   if (reset = '1') then exit reset_loop; end if;
57    unsorted_reg(ram_index) := "0000000";
58  end loop; -- ram_init
59  index2 := 0;
60  temp1 := "0000000";
61  temp2 := "0000000";
62       wait until clk = '1' and clk'event;
63  if (reset = '1') then exit reset_loop; end if;
64
65
66  MAIN_LOOP: loop
```

```
67
68  ram_load: for ram_index in 1 to N loop
69  wait until clk = '1' and clk'event;
70  if (reset = '1') then exit reset_loop; end if;
71   unsorted_reg(ram_index) := unsorted_data;
72  end loop; -- ram_load
73
74   loop1: for index1 in 1 to N-1 loop
75     index2 := N;
76     loop2: while (index2 > index1) loop
77       temp1 := unsorted_reg(index2 - 1);
78 temp2 := unsorted_reg(index2);
79 if (temp1 > temp2) then
80  unsorted_reg(index2 - 1) := temp2;
81  unsorted_reg(index2) := temp1;
82 end if;
83 index2 := index2 - 1;
84     end loop; -- loop2
85    end loop; -- loop1
86
87    ram_read : for read_index in 1 to N loop
88      sorted_data <= unsorted_reg(read_index);
89      wait until clk = '1' and clk'event;
90      if (reset = '1') then exit reset_loop; end if;
91    end loop; -- ram_read
92
93   end loop; -- MAIN_LOOP
94
95end loop; -- reset_loop
96
97   end process TOP_LOOP;
98
99  end behv;
```

**Verilog Code**

```
// Implementing the bubble sort algorithm in Verilog
// Internal RAM macro to hold the data to be sorted

module bubble_sort (unsorted_data, clk, reset, sorted_data);

parameter N = 10;
input clk;
input reset;
input [7:0] unsorted_data;
output [7:0] sorted_data;

reg [7:0] sorted_data;
reg [3:0] i;
reg [3:0] j;
reg [3:0] index1;
reg [3:0] index2;
reg [7:0] temp1;
reg [7:0] temp2;

always begin: TOP_LOOP
reg [7:0] unsorted_reg[1:N];
/* synopsys
  resource RAM_A:
variables = "unsorted_reg",
map_to_module = "DW03_ram1_s_d";
*/
  begin: RESET_LOOP
/* synopsys
      resource rlo1:
      dont_unroll = "RAM_init";
*/
// Reset all internal variables and output ports
sorted_data <= 0;
// Initializing RAM to all 0s during reset
for(i = 1; i <= N; i = i + 1) begin: RAM_init
  unsorted_reg[i] = 0;
  @(posedge clk); if (reset) disable RESET_LOOP;
end
```

```
j = 0;
index1 = 0;
index2 = 0;
temp1 = 0;
temp2 = 0;

@(posedge clk); if (reset) disable RESET_LOOP;
forever begin: MAIN_LOOP
/* synopsys
      resource rlo2:
      dont_unroll = "RAM_load";
*/
/* synopsys
      resource rlo3:
      dont_unroll = "loop1";
*/
/* synopsys
      resource rlo4:
      dont_unroll = "RAM_read";
*/
// Read data into RAM
for(i = 1; i <= N; i = i + 1) begin: RAM_load
  @(posedge clk); if (reset) disable RESET_LOOP;
  unsorted_reg[i] = unsorted_data;
end
for(index1 = 1; index1 <= N-1; index1 = index1 + 1) begin: loop1
  index2 = N;
  while (index2 > index1) begin: loop2
temp1 = unsorted_reg[index2 - 1];
temp2 = unsorted_reg[index2];
if (temp1 > temp2) begin
  unsorted_reg[index2 - 1] = temp2;
  unsorted_reg[index2] = temp1;
end
index2 = index2 - 1;
  end // loop2
end // loop1
for (j = 1; j <= N; j = j + 1) begin: RAM_read
  sorted_data <= unsorted_reg[j];
```

```
  @(posedge clk); if (reset) disable RESET_LOOP;
end // RAM_read
end // MAIN_LOOP
 end // RESET_LOOP
end // TOP_LOOP
endmodule
```

**Example 11.4 Script for Synopsys Behavioral Compiler**

```
analyze -f vhdl {pack.vhd BuS_ram.vhd}
elaborate -s bubble_sort
create_clock -period 10 clk
bc_check_design -io super
bc_time_design
schedule -io super
report_schedule -operations -var > BuS_super_10.rpt
write -hier -out bubble_sort_rtl.db
vhdlout_use_packages ={"IEEE.std_logic_1164.all; use IEEE.std_logic_arith.all;
library GTECH; use GTECH.GTECH_COMPONENTS.all; use work.TC.all"}
vhdlout_architecture_name = "BuS_RTL_ARCH"
vhdlout_dont_write_types = true
vhdlout_equations = true
write -hier -f vhdl -out bubble_sort_rtl.vhd

/* Logic Synthesis Steps*/
read -f db bubble_sort_rtl.db
compile
```

## 11.6 Behavioral Compiler Reports

In this section, we discuss the reports generated by BC for a *scheduled* design. A small portion of the report has been shown for the sake of both clarity and brevity. The reports generated provide the following details:

1. Resources used in the design. The resources used are of two types, register resources and operation resources.
2. Clock cycles in which the above resources were used.
3. Description of the Operations performed on the corresponding resource in those cycles.

### Example 11.5 Reports Generated by BC

```
NOTE: ALL REPORTS ARE GENERATED FROM VHDL CODE
****************************************************************************
          Design   : bubble_sort
****************************************************************************

*****************************************************
* Operation schedule of process TOP_LOOP: *
*****************************************************
Resource types
=====================================
  loop......loop boundaries
  p0........7-bit input port unsorted_data
  p1........7-bit registered output port sorted_data
  r44.......(7_7->1_1)-bit DW01_cmp2
  r53.......(4->4)-bit DW01_inc
  r54.......(4->4)-bit DW01_dec
  r88.......(7_4->7)-bit DW03_ram1_s_d


-------+------+-----+-------+-------+-------+-----+-----
 cycle | loop | p0  | r44   | r54   | r53   | r88 | p1
--------------------------------------------------------
....
....

  10   |......|.....|.......|.o869..|.......|.oz..|.....
  11   |......|.....|.o1546.|.......|.......|.oA..|.....
....
....

  o869.......(4_2->4)-bit SUB_TC_OP reset_loop/MAIN_LOOP/loop1/loop2/sub_77

  oz.........(4_0->7)-bit MEM_READ_SEQ_OP_read_b_DW03_ram1_s_d_state_0
reset_loop/MAIN_LOOP/loop1/loop2/MEM_READ_RAM_A_78/seq_cell_8

  o1546.....(7_7->1)-bit LT_UNS_OP reset_loop/MAIN_LOOP/loop1/loop2/gt_79
```

Consider the above reports with regard to the VHDL description of the bubble sort algorithm. Operation o869 uses the resource r54 in the cycle 10. This corresponds to the decrement operation on line 77 of the VHDL code. Operation "oz" corresponds to the memory read on line 78.

## 11.7 Is Behavioral Synthesis Right For You?

A closer look at behavioral synthesis will reveal the clear advantages over conventional logic synthesis, especially for designers accustomed to logic synthesis. However, the issue of whether it is appropriate for ones' design style, must be carefully assessed.

Behavioral synthesis requires a particular coding style to be followed. On the other hand, it can help explore several architectures by changing constraints but with minimal or no change to the source HDL. However, an efficient algorithm to start with is required to guide behavioral synthesis to an efficient implementation. Further, behavioral synthesis requires designs to be fully synchronous. The coding style required for behavioral synthesis has a significant impact on the simulation methodology. Simulation of code written for behavioral synthesis is much faster than conventional RTL simulation. In general, behavioral synthesis is suited to designs involving complex data flow or I/O operations, and several memory accesses.

Prior to adopting behavioral synthesis one must consider several issues. Think about the learning curve you'll face based on the software's appropriate coding style, tool usage, training, and your applications. The learning curve can be long depending on your current level of logic-synthesis expertise. If you are familiar with HDL coding and logic synthesis, the transition to behavioral synthesis should be relatively painless.

## 11.8 Classic Scenarios

### Scenario 1

You wish to infer a two-port RAM from your ASIC vendor's library. Both the ports of the RAM are clocked by different clocks. Is this possible using the Behavioral Compiler?

### Solution

Behavioral Compiler supports only using one clock within a process/always statement.

### Scenario 2

You are using BC to loop pipeline the following code segment. BC is unable to pipeline this loop due to data dependency of variable "mem" from one iteration to the next.

### VHDL Code

```
loop
begin
  a := mem(index);
  some operations
  mem(index) := b;
  index := index + 1;
end loop;
```

### Solution

In this design you are accessing different address of variable"mem" in each iteration.Hence, there should be no data conflict due to loop pipelining. One way to achieve this is by mapping the variable "mem" to a RAM cell and using the BC variable "ignore_memory_loop_precedences" to remove memory access precedence relations in pipelined loops. If you can't map "mem" to a RAM cell, then you must split up the array into individual variables.

## Recommended Further Readings


1. Behavioral Compiler Reference Manual
2. David Black, *Designing A 100k gate Set-Top Box ASIC using BC*, SNUG 95
3. Taher Abbasi, Pran Kurup. *Behavioral Synthesis speeds designs, reduces complexity*, EDN-Asia, May 1996.

# Appendix A

## A.1 Sample dc_shell Scripts

This appendix shows several useful dc_shell scripts.

### Example 1

A small simple script to count the total number of instances in a hierarchical design.

```
count = 0
cell_list = find (cell -hierarchy, "*")
foreach (cells, cell_list){
      count = count + 1
  }
echo "The total number of instances in this design = "  count
```

The above script can also be written as follows:

```
count = 0
foreach (cells, find(cell -hierarchy, "*")){
       count = count + 1
}
echo "The total number of instances in this design = "  count
```

### Example 2

This is a generic script to characterize, and write script for a hierarchical design. The output of the *write_script* for each instance is written to a file with the same name as the instance name. This script finds all the references in the current design, then looks for designs of the same name, then for each design, it looks for instances in the current design, then it characterizes each instance, set *current_design* to the corresponding design, then does a write script to a filename same as that of instance name. The script when run as is, will issue an error message *UID-109*, overtime DC looks for a design which is a lib cell and not a design. This can be ignored. To avoid this error remove the comment from the line *suppress_errors*. This has been commented to prevent any designs from being missed out if not available, but required in the design. Set the *current_design* to the top level design, in this case *top*.

```
current_design = top
ref_list = find(reference, "*")
file_name = ""
inst = ""
foreach (refs, ref_list){
/*suppress_errors = {UID-109} */

find(design, refs)
     if (dc_shell_status != {}) {
        filter find(cell, "*") "@ref_name == refs"
        inst = dc_shell_status
          foreach(one, inst){
             file_name = one + ".scr"
             current_design top
             characterize one
             current_design  = refs
             write_script > file_name
             compile
             report_timing >> file_name
        }
   } else {
        current_design top
   }
}

current_design top
report_timingreport_timing > timing_report_top
```

## Example 3

This script groups all the instances whose references are from a certain technology library into a separate block or sub-design. This or modifications of this script can come in handy when one is dealing with designs which have cells from different libraries.

```
ref_list = find(reference, "*")
inst = ""

foreach (ref, ref_list){
new_ref = "libB/" + ref
foreach (rr, new_ref){
   lib-cells = find(cell, rr)

foreach(one, lib-cells) {
      filter find(cell, "*") "@ref_name == ref"
     inst = dc_shell_status + inst
    }
  }
}

group  inst -design_name new_block
echo "The following instances are being grouped to create a new design called
new_block"
echo inst
```

## Example 4

At the top level, if one has a number of cells, a few sub-designs, and others, library cells, *report_cell* is one way to get the entire list of cells and their instance names at the top level. This script helps to get the instance names of only the subdesigns in the hierarchy.

```
current_design = top
design_list = find(design, "*")
inst = ""
foreach (des, design_list){
      filter find(cell, "*") "@ref_name == des"
      inst = dc_shell_status + inst
foreach (inst_name, inst){
```

```
    echo inst_name +  " is an instance of " + des
  }
}
```

## Example 5

This script does the following to modify the Synopsys library for in place optimization.

1. Reduces area of low drive cells by 10
2. Divides the capacitance and fanout load on pins by 1.6.
3. Checks to see if all the low drive cells have footprints.

Variables that can be specified.

1. cell_list (The list of low drive cells)
2. fudge_area (This is set to 10)
3. fudge (This is set to 1.6)

If you remove the library from DC memory for some reason, be sure to include this script again. This will ensure that you are using the library for in place optimization.

```
read -f db <your_technology_library>
/* This part of the script modifies the area of the cells by 10, this value can be changed
by setting fudge_area to another value cell_list is the list of cells for which area is being
reduced. To add cells to this list add them to the cell_list */
fudge_area = 10.0
area_new = 0
area_old = 0
cell_list = ""
area_count = 1
original_area_list = ""
cell_list = {"LIBA/and2l LIBA/and3l LIBA/ao21l LIBA/ao22l LIBA/aoi21l \
LIBA/fd00p1al LIBA/fd01p1al LIBA/invl LIBA/nd2l LIBA/nd3l LIBA/nr2l \
LIBA/nr3l LIBA/oa21l LIBA/oa22l \
LIBA/oai21l LIBA/oai22l LIBA/or2l LIBA/xnor2l LIBA/xor2l"}
get_attribute findfind(cell, cell_list) area
original_area_list = dc_shell_status
foreach(area_old, original_area_list) {
cell_count = 0
foreach(lp-cell, cell_list){
    cell_count = cell_count + 1
    if (area_count == cell_count) {
```

```
echo "Old area for "+ lp-cell + " is " +  area_old
area_new   = area_old - fudge_area
echo " New Area for " + lp-cell + " is " + area_new
set_attribute find(cell, lp-cell) area area_new
} else {
}
}
        area_count = area_count + 1
}
get_attribute find(cell, cell_list) area
new_area_list = dc_shell_status
```

Alternatively, one can specify areas on the cells equal to their max_fanout. Because, higher drive cells have a higher max fanout than low drive cells, one can specify an area equal to the max fanout of the output pin of that cell. This will avoid having to find the appropriate area fudge factor

```
pin_list = ""
max = ""
pin_name = ""
ccc = ""
ccc = find(cell, "LIBA/*")
new_ppp = ""
echo ccc
foreach(clist, ccc) {
     pin_name = clist + "/*"
     echo pin_name
     pin_list = find(pin, pin_name)

     foreach(ppp, pin_list){
               new_ppp = clist + "/" + ppp
               get_attribute find(pin, new_ppp) pin_direction
          if (dc_shell_status == out){
               get_attribute find(pin, new_ppp) max_fanout
               max = dc_shell_status
               echo "max-fn = " max
          foreach (new_area, max){
               set_attribute clist area  new_area
               }
```

```
            }
            }
        }

/* The part below changes the capacitance and fanout_load on the pins */
/* If fudge factor is different from 1.6 then simply change fudge */
fudge = 1.60
pin_list = ""
fan = ""
cap = ""
new_cap = 0.00
new_fan = 0.00
pin_name = ""
ccc = ""
new_ppp = ""
ccc = find(cell, "LIBA/*")
foreach(clist, ccc) {
        pin_name = clist + "/*"
        echo pin_name
        pin_list = find(pin, pin_name)
foreach(ppp, pin_list){
      new_ppp = clist + "/" + ppp
      get_attribute find(pin, new_ppp) pin_direction
if (dc_shell_status == in){
      get_attribute find(pin, new_ppp) capacitance
      cap = dc_shell_status
      echo "capacitance of " new_ppp " = " cap

get_attribute find(pin, new_ppp) fanout_load
      fan = dc_shell_status
      echo "Fanout load of " new_ppp " = " fan
foreach(cap_value, cap){
      new_cap = (cap_value)/(fudge)
      echo "New cap of " new_ppp " = " new_cap
}
foreach(fan_value, fan){
     new_fan = (fan_value)/(fudge)

echo "New fanout load of " new_ppp " = " new_fan
```

```
}

set_attribute find(pin, new_ppp) capacitance new_cap

set_attribute find(pin, new_ppp) fanout_load new_fan
      }
   }
}

/* This part of the script will check for low power cells which are missing footprints
cells_without_footprint must return an empty "", if not footprints are missing for some of
the low power cells*/

cells_without_footprint = ""

foreach (cells, cell_list) {
get_attribute find(cell, cells) cell_footprint
if (dc_shell_status = 0) {
      echo "Footprint is missing for " cells
      cells_without_footprint = cells_without_footprint + " " + cells
     }
}

echo cells_without_footprint
echo "The Original areas are as follows "
echo  original_area_list
echo "The Modified areas are as follows"
echo  new_area_list
```

## Example 6

For in place optimization, often back annotated information after P&R is in the form of *set_load* script. After including these set load information, it might be required to find all the nets which have load values set and all the nets which do not. Then for all those without load values, it might be required to set specific load value. The following script can be used to do this.

```
nets_list = find(net -hierarchy, "*")
net_count_with_load = 0
net_names_without_load = " "
net_count_without_load = 0
```

```
foreach(nets, nets_list) {
   get_attribute find(net, nets) load
    if (dc_shell_status != {}) {
        net_count_with_load = net_count_with_load + 1
     } else {
        net_count_without_load = net_count_without_load + 1
     if (net_count_without_load > 1) {
        net_names_without_load = net_names_without_load + "," + nets + ", "
     } else {
        net_names_without_load =  nets
      }
    }
}

echo net_count_with_load + " nets have loads set on them"
echo net_count_without_load + " nets do not have loads set on them"
echo net_names_without_load
foreach (without_load, net_names_without_load) {
       set_load 3.0 find(net, without_load)
}
```

## Example 7

At the very top level of a hierarchical design, it is often required to ungroup all the sub-levels in the design. The *ungroup -flatten -all* command recursively ungroups all sub-levels provided all the sub levels in the design are not dont_touched. This script traverses the hierarchy and removes the dont touch on all sub designs and then ungroups all at the top.

```
des_list = find(reference,"*")
foreach (des, des_list)
              {
     suppress_errors = {UID-109}
     find(design, des)
       if (dc_shell_status != {}) {
           current_design = des
remove_attribute find(design,des) dont_touchremove_attribute find(cell -hierarchy, "*")
dont_touch
} else {
 suppress_errors = {}
}
```

```
}
current_design = top
ungroup -flatten -all -simple_names
```

### Example 8

You wish to find the number of flip flops in the design, the number of latches in the design, the number of latches/flops in a particular sub-module, A, of the design.

```
cell_list = find(cell -hier, "*")
count0 = 0 ;
foreach (cells, cell_list)
     {
           count0 = count0 + 1 ;
     }
echo "The total number of instances in this design =" count0

all_registers -edge_triggered
flop_list = dc_shell_status
count = 0 ;
count2 = 0 ;
count3 = 0 ;

foreach (cells, flop_list){
                count = count + 1
                }
echo "The total number of flops in this design = " count
all_registers -level_sensitive
latch_list = dc_shell_status
foreach (cells, latch_list) {
           count2 = count2 + 1
     }
echo "The total number of latches in this design = " count2
current_design = A
all_registers -edge_triggered
dp_flop_list = dc_shell_status
foreach (cells, dp_flop_list){
                count3 = count3 + 1
                }

echo "The total number of instances in this design = " count0
```

```
echo "The total number of flops in this design = "  count
echo "The total number of latches in this design = "  count2
echo "The total number of flops in A = " count3
```

### Example 9

dc_shell script to find all nets connected to three-state pins in the design.

```
cell_list = {}
pin_list = {}
refname = {}
temp = ""
rname = ""
find(cell, "*")
cell_list = dc_shell_status
foreach(cname, cell_list){
  get_attribute cname ref_name
  foreach(rname, dc_shell_status){}
  find(pin, "lsi_10k/" + rname + "/*")
  pin_list = dc_shell_status
  get_attribute pin_list three_state
  if (dc_shell_status == {}){
echo "This is not a three state cell";
  } else {
echo "This is a three-state cell"
      filter pin_list "@pin_direction == in"
      pin_list = pin_list - dc_shell_status
filter pin_list "@three_state == {}"
pin_list = pin_list - dc_shell_status
    foreach(pname, pin_list){}
temp = cname + "/" + pname
all_connected temp
      foreach(temp1, dc_shell_status){}
echo "Net connected to 3-state pin" temp "is" temp1 >> file_name
  }
}
```

## A.2 Sample Synopsys Technology Library

```
library (my_lib){
 date : "February 24, 1992";
 revision : 2.2;
cell(AN2) {
  area : 2;
  pin(A) {
    direction : input;
    capacitance : 1;
  }
  pin(B) {
    direction : input;
    capacitance : 1;
  }
  pin(Z) {
    direction : output;
    function : "A B";
    timing() {
      intrinsic_rise : 1.00;
      intrinsic_fall : 1.00;
      rise_resistance : 1.00;
      fall_resistance : 1.00;
      slope_rise : 1.0;
      slope_fall : 1.0;
      related_pin : "A";
    }
    timing() {
      intrinsic_rise : 1.00;
      intrinsic_fall : 1.00;
      rise_resistance : 1.00;
      fall_resistance : 1.00;
      slope_rise : 1.0;
      slope_fall : 1.0;
      related_pin : "B";
    }
  }
}
cell(OR2) {
  area : 2;
```

```
  pin(A) {
    direction : input;
    capacitance : 1;
  }
  pin(B) {
    direction : input;
    capacitance : 1;
  }
  pin(Z) {
    direction : output;
    function : "A+B";
    timing() {
      intrinsic_rise : 1.00;
      intrinsic_fall : 1.00;
      rise_resistance : 1.00;
      fall_resistance : 1.00;
      slope_rise : 1.0;
      slope_fall : 1.0;
      related_pin : "A";
    }
    timing() {
      intrinsic_rise : 1.00;
      intrinsic_fall : 1.00;
      rise_resistance : 1.00;
      fall_resistance : 1.00;
      slope_rise : 1.0;
      slope_fall : 1.0;
      related_pin : "B";
    }
  }
}
cell(IV) {
  area : 1;
  pin(A) {
    direction : input;
    capacitance : 1;
  }
  pin(Z) {
    direction : output;
```

```
    function : "A'";
    timing() {
     intrinsic_rise : 1.00;
     intrinsic_fall : 1.00;
     rise_resistance : 1.00;
     fall_resistance : 1.00;
     slope_rise : 1.0;
     slope_fall : 1.0;
     related_pin : "A";
    }
  }
}
cell(VCC) {
  area : 1;
  pin(Z) {
    direction : output;
    function : "1";
  }
}
cell(GND) {
  area : 1;
  pin(Z) {
    direction : output;
    function : "0";
  }
}
}
```

# A.3 Sample Synopsys Technology RAM Library Model

```
cell (sample) {
 area : 60    /* area of cell */
 pin(A1) {
   direction : input
   capacitance : 0.8   /*  capacitance of pin  */
   fanout_load : 1.0   /*  fanout load   */
 }
 pin(A2) {
   direction : input
   capacitance : 0.8
   fanout_load : 1.0
 }
 pin (Z1) {
   direction : output
   max_fanout : 10
   timing() {
     intrinsic_rise : 1.490000   /*  intrinsic delay of A1 to Z1  */
     intrinsic_fall : 1.800000
     rise_resistance : 0.185000  /*  drive strength of output port */
     fall_resistance : 0.059000
     related_pin : "A1"
 }
  timing() {
     intrinsic_rise : 1.590000
     intrinsic_fall : 1.700000
     rise_resistance : 0.185000
     fall_resistance : 0.059000
     related_pin : "A2"
 }

}
 pin (Z2) {
   direction : output
   max_fanout : 10
   timing() {
     intrinsic_rise : 1.490000
     intrinsic_fall : 1.800000
     rise_resistance : 0.185000
```

```
      fall_resistance : 0.059000
      related_pin : "A1"
    }
    timing() {
      intrinsic_rise : 1.690000
      intrinsic_fall : 1.900000
      rise_resistance : 0.185000
     fall_resistance : 0.059000
      related_pin : "A2"
    }

  }
 }
}
```

## References

1. *Design Compiler Reference Manual.*
2. *VHDL Compiler Reference Manual*
3. *A Guide to VHDL*, Stanley Mazor, Patricia Langstraat.
4. *Synopsys Methodology Notes.*
5. *ASIC & EDA, Feb 1994.*
6. *VHDL: Hardware Description and Design*, Lipsett, Schaefer, Ussery
7. *Computer Design, Oct 94.*
8. *The Verilog Hardware Description Language* - Donald Thomas, Philip Moorby
9. *Field Programmable Gate Arrays* - Stephen D. Brown, Robert J. Francis, Jonathan Rose, Zvonko G. Vranesic

# *Index*

## A



## B



## C

## D



## E



## F



## G



## H



## I

**J**



**L**



**M**



**N**



**O**



**P**



**R**

## S

**T**



**U**



**V**



**W**



**X**